The Concept of Reduction

PHILOSOPHICAL STUDIES SERIES

VOLUME 121

For further volumes:
http://www.springer.com/series/6459

Raphael van Riel

The Concept of Reduction

Springer

Raphael van Riel
Institut für Philosophie
Universität Duisburg-Essen
Essen, Germany

ISBN 978-3-319-34552-9 ISBN 978-3-319-04162-9 (eBook)
DOI 10.1007/978-3-319-04162-9
Springer Cham Heidelberg New York Dordrecht London

Softcover reprint of the hardcover 1st edition 2014

Printed on acid-free paper

Springer is part of Springer Science+Business Media (www.springer.com)

For Katharina

Preface

This book is a revised version of my doctoral thesis submitted at the Ruhr-University of Bochum in July 2010. The thesis was written under the supervision of Albert Newen and Stephan Hartmann. The core arguments were discussed with my colleagues in Bochum and at the Tilburg Center for Logic and Philosophy of Science (TiLPS), where I spent a couple of months in 2009 as a visiting fellow. Several papers emerged from this thesis, brief summaries of which are included in the present book. I started working on the manuscript again in 2012, while spending two wonderful months in Barcelona as a visiting fellow in the LOGOS group.

First of all, I wish to thank my supervisors Albert Newen and Stephan Hartmann for their helpful comments on the manuscript and the long discussions we had on this topic. I am grateful for their guidance. I would like to thank Markus Werning and Ulrich Pardey, who were on the committee and provided helpful feedback. A somewhat independent 'thank you' goes, again, to Albert Newen for his strategic and motivational support as my boss for more than 5 years at the University of Bochum. The time was great. I also wish to thank my former colleagues for helpful discussions, for reading sections of the manuscript, and for providing distraction when needed: Leon de Bruin, Eva Maria Jung, Lena Kästner, Ulrike Pompe, Tobias Schlicht, Tomoo Ueda, Anna Welpinghus, and, especially, Markus Eronen, Christoph Michel, Tobias Starzak, and Gottfried Vosgerau. I had extremely helpful discussions with Alexander Reutlinger, Christian Sachse, Patrice Soom, and Jan Sprenger. I received valuable comments from a group of philosophers at Brown University, and I would like to thank Douglas Kutach for his feedback on a paper that summarized an early version of the core argument of Chap. 3. I would also like to thank Gerhard Schurz for inviting me to his colloquium, where I received extremely valuable comments. I am grateful to Dan López de Sa and Moritz Schulz for inviting me to the PERSP-Metaphysics seminar in Barcelona, where a paper of mine was intensively discussed – a paper significant portions of which later entered Chaps. 3 and 9. Working with Robert Van Gulick on the entry 'Scientific Reduction' for the *Stanford Encyclopedia of Philosophy* helped me to get clearer about numerous subtleties of the reduction debate – I wish to thank him for that. I received valuable input during workshops and in colloquia in Lausanne, Milan,

Krakow, Helsinki, Tilburg, and Geneva. I am especially grateful for the intense, often nightlong discussions I had with Heiner Koch, Nick Haverkamp and Miguel Hoeltje, who also read and commented on significant portions of this book. I also wish to thank Daniel James for reading parts of the book and for intense discussions. An anonymous referee provided extremely helpful comments not only on particular sections, but also on the book's overall structure – in the light of these comments I restructured the entire book; I truly hope the result does justice to the referee's valuable suggestions, at least to some extent.

I would also like to thank the foundations that enabled me to conduct research, stay abroad and travel to conferences: the *Studienstiftung des deutschen Volkes* and the *Mercator Stiftung*. The final version of this manuscript was produced with the support of a generous *Dilthey-Fellowship* of the *Volkswagen Foundation*. Thanks also to Ties Nijssen and Christi Lue from *Springer* for their support.

Finally, I would like to express my gratitude to my former teachers at the University of Hamburg, especially Harald Wohlrapp and Wolfgang Künne, and to my family, who encouraged me to pursue a philosophical career, and who enabled me to carry out this project – my parents, Peter and Ulrike, with great affection, my brother and sister, and Hartmut Meißner. My deepest thanks go to my wife, Katharina, for being patient, for always putting things into perspective, and for her encouragement. And thanks to my children – for everything.

Essen, Germany Raphael van Riel

Contents

Chapter 1
Introduction

This book is about the notion of reduction, about the question of how an appropriate explication of this notion should look like, and about its history, i.e. its role especially in the philosophy of science and in the philosophy of mind. The following slogan gives a first, rough idea of how the term 'reduction' will be understood here: *The concept of reduction is supposed to reconcile diversity and directionality with unity, without relying on elimination.* If *water* reduces to H_2O then there is just one thing rather than two – thus we get *unity. Unity* does not come at the price of *elimination* – claiming that water reduces to H_2O, we do not thereby claim that there is no water. But what about diversity and directionality?[1] Intuitively, there should be a difference between water and H_2O, such that we get diversity. This is required for there to be directionality: in a sense, if water reduces to H_2O, then H_2O is prior to, or more basic than water. At least, if water reduces to H_2O, then H_2O does not reduce to water. And this is where trouble starts:

If water reduces to H_2O, such that there is unity, how can there be a difference between water and H_2O? On a strong interpretation of 'unity', if there is unity between water and H_2O, then there is just one thing. And then, there can be no difference between water and H_2O, and, hence, no directionality. Thus, it is not easy to see how to account for unity and diversity at the same time. This is a problem any appropriate explication of the notion of reduction has to solve. Three candidate strategies suggest themselves.

If we interpret 'unity' more liberally, we may introduce relations weaker than that of identity, relations such as supervenience or ontological dependence. On this interpretation, we get some sort of (weaker) unity, and at the same time we get an idea of how there can be diversity as well as directionality. Although mental

[1]The reader might wonder why I talk about 'directionality' which is, contrary to 'asymmetry', not a technical term; this is the very reason for why I prefer using it at the moment. Later, it will be argued that reduction is not an asymmetric relation, despite the fact that if an appropriate instance of '*a* reduces to *b*' expresses a truth, then the corresponding instance of '*b* reduces to *a*' expresses a falsehood. The reasons will become clear below.

R. van Riel, *The Concept of Reduction*, Philosophical Studies Series 121,
DOI 10.1007/978-3-319-04162-9_1, © Springer International Publishing Switzerland 2014

processes might *supervene* or *ontologically depend* upon physical processes, and, hence, be different from the latter, we still have unity in the sense that there is one level that fixes all the rest. What fixes all the rest is, in this very sense, *prior* to the rest, such that the idea of directionality is, so to speak, built into our notion of dependence right from the start.

Alternatively, we may stick to the strong reading of unity in this sense: there is unity between x and y iff $x = y$, and there is unity between *Fs* and *Gs* iff for every x that is F, there is a y that is G, such that $x = y$. In this case, we must look for diversity and directionality somewhere else. In the philosophy of science, it is common to discuss reduction in terms of a relation that primarily holds between theories or scientific models. Now, for one theory to reduce to another theory, these theories should better be distinct. Even if there is no difference between water and H_2O, there surely is a difference between our folk-theory of water and our chemical theory of H_2O. And it is not hard to see why we should regard the latter as more basic than the former. Thus, on this interpretation, we get diversity and directionality at the level of theories and we still may have strong unity at the level of properties or worldly objects.

Finally, for the eliminativist, unity, diversity and directionality come very cheap. If you take an eliminativist stance regarding, say, mental properties, you commit yourself to the idea that there is diversity in the field of *allegedly* existing entities (mental entities in addition to neural entities), and that there is unity regarding the *actually* existing entities (only neural entities), and that there is directionality in the sense that our ontologically appropriate talk (neural talk) is “more basic” than our ontologically misguided talk (psychological talk).

This is not a book on the notion of elimination, nor on more liberal notions such as supervenience or ontological dependence. Philosophers who adopted some sort or other of supervenience-theory of the mind usually do not regard themselves as reductionists, and rightly so. Similarly, metaphysicians who work on relations of ontological dependence do not usually cast their theses in terms of *reduction*. Reviewing the reduction literature, it will turn out that the use of the term ‘reduction’ across the philosophical sub-disciplines suggests a strong interpretation of the kind of unity that goes together with reduction. So, let me give a slightly more precise version of the slogan introduced above: *The concept of reduction is supposed to reconcile diversity and directionality with* strong *unity.* Strong unity is the unity of identity.

This is the concept this book is about. Now, should we, in the light of what was said so far, assume that this concept applies to theories, rather than properties, processes or individuals? This way, we could easily eschew the problem just sketched. A large part of this book deals with this question, and it will be argued that notions of theory-reduction are parasitic on a notion of “ontological” reduction, or a notion of reduction that applies primarily to properties, facts, kinds, individuals, or processes. If this is correct, it seems that there is no easy way out of the problem sketched above. A careful investigation of this problem will, however, lead to a detailed and fruitful explication of the notion of reduction. So, let me state it more precisely.

1.1 A Puzzle About Reduction

Let us assume that there is a reading of sentences of the form '*F-ness* reduces to *G-ness*' (where '*F-ness*' and '*G-ness*' are placeholders that stand for terms referring to kinds) under which they express a truth only if a corresponding sentence of the form '*F-ness* = *G-ness*' expresses a truth as well. This idea generates a puzzle. Consider again our example: 'Water reduces to H_2O' expresses a truth only if water is identical to H_2O (here, I suggest treating 'water' and 'H_2O' as terms designating kinds; nothing hinges on this, we could also replace them by designators such as 'the property of being water'). At the same time, H_2O does not reduce to water! The reduction relation appears to be asymmetric (and hence irreflexive). Identity is not. How can this possibly be the case?

1.2 A Sketch of a Solution

Roughly, it will be argued that '_reduces to_' generates hyper-intensional contexts. This points to the fact that the truth of the sentence partly depends on the different ways in which the reduced/reducing object is presented in the sentence (in the two argument-positions of '_reduces to_'). This, in turn, enables us to account for diversity, directionality, and (what looks like, but literally is not) asymmetry: water reduces to H_2O (if it does) because (i) water just is H_2O, and (ii) the meaning of 'water', or the conceptual content expressed by 'water', presents us with water in a way that is relevantly different from the way the meaning of 'H_2O' presents us with water. Intuitively, this difference is relevant because the concept of H_2O presents us with water under *constitutive properties*, or under properties that are on a lower level of constitution than those under which the concept of water presents us with water (if there is such a concept, that is: if 'water' is not a directly referential expression).

This line of thinking can easily be connected to one of the core assumptions in the reduction literature: It is common to regard the relevant reduction-statements (or, in the case of theory-reductions: bridge-laws) as necessary *a posteriori*, and, consequently, the equivalent laws, theories and so forth as equivalent a posteriori. This sort of aposteriority is best explained by the fact that the terms flanking the reduction predicate have different conceptual contents. Building on this Fregean picture of language, according to which we have to distinguish between *conceptual content of* and *what is designated or signified by* an expression, we will be able to develop a coherent and explanatorily powerful account of the concept of reduction: *Reduction is tied to conceptual contents presenting us with the same things in different ways, as having specific properties.*

This notion of a mode of presentation will be spelled out in terms of *property-structures* – intuitively, in terms of structures of properties under which an expression's meaning or conceptual content presents us with the object the expression designates or signifies. Property-structures thus determine a specific way an entity is given in discourse. These ways differ for different sorts of discourse, and some might be, in a sense to be specified, more fundamental than others. Tying the notion of reduction to that of reductive explanation to make sense of the idea of *descriptive diversity* and *explanatory directionality* that goes together with *ontological unity*, a characterization of reduction will be given, which adequately captures one use of the term in the philosophical debate, especially in the philosophy of mind, but also in large parts of the philosophy of science.

In a nutshell, a sentence of the form '*F-ness* reduces to *G-ness*' expresses a truth if and only if (i) for every x, if x is F then (x is F *because* x is G), and (ii) $F\text{-}ness = G\text{-}ness$. Similarly, a sentence of the form 'ϕ-*ing* reduces to ψ-*ing*' expresses a truth if and only if (i) for every x, if x ϕ-*s* then x ϕ-s by ψ-*ing*, and (ii) $\phi\text{-}ing = \psi\text{-}ing$. In order to exclude cases of *conceptual analysis*, one may want to add that the concept of an F (a ϕ-*ing*) is not analyzable in terms of the concept of a G (a ψ-*ing*).

The explication of the notion of reduction proposed here can be used to illuminate aspects regarding descriptive pluralism in worlds with a monistic ontology, or, similarly, regarding descriptive pluralism for *parts* of worlds with a monistic ontology. Being an explication of a notion of reduction, which guarantees ontological unification by *identification*, it is an explication of a specific concept of *identity-based reduction*. Derivative notions, such as a notion of theory reduction, of token-reduction, of partial reduction, of fact-reduction and of plural reduction, as well as a notion that enables us to form generic reduction statements can be explicated. Moreover, the so explicated notion sheds light on a number of related topics, such as reduction and unification, reduction and mechanistic explanation, reduction and grounding, and reduction and intervention.

1.3 A Note on Methodology

The notion of reduction is usually introduced in some broader context, either as a core notion (amongst others) in the philosophy of science (Oppenheim and Putnam 1958; Nagel 1961; Stegmüller 1986), as a prerequisite to address other issues regarding the structure of some special science (Schaffner 1993), or within the context of issues regarding some version or other of reductionism or antireductionism (Sarkar 1998; Bickle 1998, 2003; Feigl 1967; Kim 1992; Chalmers 1996). In these contexts, discussions of previous models of reduction do not loom large, and several related issues, which are not the target of this book, like mere *replacement*, reductionist *procedures* and the *epistemology* of reductions, are all considered to be part of a discussion of the concept of reduction. In this book, a slightly different strategy is adopted. 'Reduction' is an expression that plays a distinctive role in a number of explanatory contexts. This role imposes certain constraints on how this notion should be understood. In this book, a definition is proposed, which captures this role by paying attention to these constraints. Roughly, the method can be understood as a cognate of Carnapian *explication* (more on this in Chap. 2). Thereby, the target of this book differs from most of the targets of other publications on reduction; frequently, these comprise a discussion of reduction*ism*, and not only a discussion of the *notion of reduction*. So, how does the explication offered here relate to debates about reductionism?

1.4 Reduction and Reductionism

Offering an explication of the concept of reduction, we define a concept that is used in formulations of a variety of ontological positions, such as reductionism, non-reductive physicalism, unificationist positions in the philosophy of science, and the like. Versions of *general* reductionism underlie orthodox empiricist pictures of unification. In its domain-specific form, i.e. in a form which concerns some fragments of (scientific) discourse only, reductionism has undergone some sort of revival in the past decades, due to relatively recent developments in the neurosciences, which were enthusiastically described as pushing us towards a reductive science of the mind. In addition, the notion of reduction addressed here underlies the debate between type- and token-identity theorists, and it is crucial for understanding fundamental aspects of *models* of reduction proposed in the literature, in the philosophy of science as well as in the philosophy of mind. Offering an explication of this concept, this book has got little to say about whether or not reductionism – either some general version according to which everything reduces to some fundamental level, or some domain-specific version according to which at least some high-level items reduce to some lower level ones – actually holds. It will be assumed that some variant of reductionism is conceptually possible and not obviously misguided – be it domain-specific or general. Whereas arguments in favor of or against reductionism heavily depend upon assumptions about how the actual world is shaped, or how successful science seems to be, and while such arguments have an impact on how to describe scientific findings and progress, the aim of this book is more modest: it aims at *conceptual* clarifications. It is thus not concerned with issues regarding the relation between epistemological and ontological versions of reduction*ism*, which are not to be confused with issues concerning epistemological and ontological conceptions of reduction: The former[2] concern issues regarding the accessibility of reductions in the actual world, the latter are concerned with conceptual issues of shaping the notion of reduction. However, giving a tenable explication of the notion of reduction that makes the appearance of paradox vanish bears upon the plausibility of reductionist claims; once we have come to see what these claims actually consist in, or how they should best be construed, we will be in a position to judge these

[2]This is the main issue between those who oppose the idea that actually, we could carry out a unification of sciences by way of actual steps of reduction, and those who take this assumption, at least as an assumption guiding the investigation, to be promising. The former view is discussed at some length in Darden and Maull (1977), a prominent example of the latter is Oppenheim and Putnam (1958).

claims more appropriately. Solving some puzzles about the notion of reduction we thus solve some puzzles about the idea of reductionism. The book aims at a fruitful explication of what reductionists, in virtue of being reductionists, are committed to. Moreover, this book's historical parts reconstruct the different suggestions for defining the notion of reduction in the debate on reductionism. In this limited sense, this book is concerned with reduction*ism*.

1.5 The Book's Structure

This book consists of two parts, with four chapters each, and a short concluding chapter. It starts with an explication of the notion of reduction as conceived of here, building on a careful examination of the puzzle of reduction. The second part shows the fruitfulness of the concept so explicated. The structure can be captured by the two questions and eight theses summarized in Table 1.1. These theses do not exhaust the content, but they capture the chapters' main contributions to the overall structure of this book. This structure will be referred to in the chapters' conclusions to orient the reader.

Part I, besides offering a note on methodology, discusses the problematic components of an appropriate characterization of the concept of reduction, namely, unity (Chap. 2), diversity (Chap. 3) and directionality (Chap. 4), to offer a fruitful explication of the concept (Chaps. 4 and 5).

Chapter 2 deals with preliminary issues. It offers a taxonomy of different conceptions of reduction and closely related notions and sheds further light on the

Table 1.1 The book's structure

The book's structure	
Q1:	**How can we reconcile *diversity* and *directionality* with strong *unity*? (Part I)**
Th.1:	*Strong unity* is the unity of identity. *Method*: To answer Q1 is to come up with an *explication* of the concept of reduction (Chap. 2).
Th. 2:	*Diversity* is conceptual in nature, as a discussion of the puzzle of reduction shows. It is to be accounted for in terms of the conceptual contents of the expressions that pick out the reduced/the reducing entity in a true reduction statement (Chap. 3).
Th. 3:	*Directionality* is explanatory in nature; reductive explanation is a cognate of mechanistic explanation (Chap. 4).
Th. 4:	In this spirit, a core notion and derivative notions of reduction can be explicated; the explication fits paradigmatic cases, is coherent (solves the puzzle) and satisfies an intuitive job-description (Chaps. 4 and 5).
Q2:	**How can this explication be further motivated? (Part II)**
Th. 5:	It sheds light on the reduction debate in the philosophy of mind (Chap. 6).
Th. 6:	It sheds light on the reduction debate in the philosophy of science (Chap. 7).
Th. 7:	It is as committal as and more fundamental than rival explications (Chaps. 8 and 9).
Th. 8:	It sheds light on closely related issues, such as reduction and unification, pragmatic benefits of reduction, and notions of scientific levels (Chap. 9).

idea of *strong unity*. It discusses the methods underlying the various approaches to reduction. It is argued that the method of *explication* is the only appropriate method in the present context. A number of criteria of adequacy any appropriate explication of the notion of reduction has to meet are identified, and the relevant terminology is introduced.

Chapter 3 discusses the puzzle of reduction. It is shown that '_reduces to_' generates hyper-intensional contexts, and that in order to account for what looks like the asymmetry of identity-based reduction, the difference between the conceptual contents of the expressions that pick out the reduced and the reducing entity should be taken into account. *Diversity* is conceptual in nature.

Chapter 4 opens with a first explication of the concept of reduction, tying *directionality* to *explanatory dependence*. This characterization, however, remains silent about the details of reductive explanation. It is argued that the best candidate for a more thorough characterization of *reductive explanation* is a cognate of what is often discussed under the heading 'mechanistic explanation'. Recent models of mechanistic explanation are discussed, according to which the functioning of a complex is explained in terms of the functioning of its parts, and four core aspects of such models, which, in the present context, need substantial revision, are identified: (i) their scope (*mechanisms* only), (ii) the often built-in functional-teleological aspect (via reference to properties such as *being organized to do something*) (iii) their (problematic and ill-justified) focus on 'how'-questions, and (iv) their ontological underpinnings.

Based on revisions of these four aspects, a characterization of reductive explanation is suggested in Chap. 5. Mechanistic dependence is usually conceived of as a dependence relation between different entities, a complex and its constituents. Reductive "dependence" involves one entity and two semantic values that both represent this entity. The relation between these semantic values that gives rise to reductive dependence is a relation between modes of presentations, where modes of presentations are characterized in terms of *property structures*. Intuitively, different property structures can depict one and the same entity at different levels of constitution. A more thorough version of the definition of the core notion of reduction can be offered. Several derivative notions can be defined. *Explanatory directionality* turns out to be tied to *conceptual diversity*. Two rival explications, a two-dimensionalist and a pragmatist one, are briefly discussed.

Part II illustrates the explication's fruitfulness, and it argues that in fact, there is a common core shared in the reduction debates in the different philosophical fields that is adequately captured by the explication proposed here.

The next two chapters apply the core notion and a few derivative notions to reduction debates in the philosophy of mind (Chap. 6) and the philosophy of science (Chap. 7). Type-identity theories, token-identity theories, conceptions of non-reductive physicalism, and the model of functional reduction can easily be reconstructed based on the explication proposed here, sometimes in an improved form (for example, on the view presented here, it becomes immediately clear where the directionality stems from – an aspect that has been ignored in the heat of the debate about type-*identity* theories). Similarly, it will be shown that a derivative notion of theory reduction can be defined that captures core tenets of Nagelian reduction, and that is required to fully understand the functioning of replacement models.

Chapter 8 argues that the explication proposed here is more fundamental than notions of theory reduction. First, the explication proposed here is not more expensive, neither metaphysically nor semantically. It suggests that models of theory reduction fail, because, under specific circumstances, they cannot account for the directionality of reduction. Moreover, if we adopt the explication proposed here, we can *explain* features of theory reductions that would turn out to be *basic* otherwise.

Chapter 9 continues parts of the discussion of Chap. 8 and applies the explication proposed here to broader issues in the philosophy of science and metaphysics. It is argued that various forms of unification are to be expected given a (reductive) theory change occurs. These forms of unification can easily be characterized. It is argued that reduction, as conceived of her, is a form of grounding, and it is shown why a reduction will go together with pragmatic benefits: The range of possible intended interventions is expanded once we learn about a reduction relation.

Chapter 10 summarizes the main points.

Bibliography

Bickle, John. 1998. *Psychoneural reduction: The new wave*. Cambridge, MA: MIT Press.

Bickle, John. 2003. *Philosophy and neuroscience: A ruthlessly reductive account*. Dordrecht: Kluwer.

Chalmers, David. 1996. *The conscious mind*. Oxford: Oxford University Press.

Darden, Lindley, and Nancy Maull. 1977. Interfield theories. *Philosophy of Science* 44: 43–64.

Feigl, Herbert. 1967. *The "mental" and the "physical". The essay and a postscript*. Minneapolis: University of Minnesota Press.

Kim, Jaegwon. 1992. Multiple realization and the metaphysics of reduction. *Philosophy and Phenomenological Research* 52: 1–26.

Nagel, Ernest. 1961. *The structure of science. Problems in the logic of explanation*. New York: Harcourt, Brace & World, Inc.

Oppenheim, Paul, and Hilary Putnam. 1958. The unity of science as a working hypothesis. In *Concepts, theories, and the mind-body problem*, ed. Grover Maxwell, Herbert Feigl, and Michael Scriven, 3–36. Minneapolis: Minnesota University Press.

Sarkar, Sahotra. 1998. *Genetics and reductionism*. Cambridge: Cambridge University Press.

Schaffner, Kenneth. 1993. *Discovery and explanation in biology and medicine*. Chicago: The University of Chicago Press.

Stegmüller, Wolfgang. 1986. *Theorie und Erfahrung: Dritter Teilband. Die Entwicklung des neueren Strukturalismus seit 1973*, vol. 3. Berlin/Heidelberg/New York: Springer.

Part I
The Concept of Reduction – An Explication

Chapter 2
How to Approach Reduction: Explication and Meta-science

Reduction, in the sense of 'reduction' we are interested in, reconciles diversity and directionality with strong unity. 'Reduction', however, can be given different interpretations; it is a technical term whose meaning largely depends on the theoretical goals of those who use the term. This chapter offers a disambiguation of the term, and it addresses methodological issues regarding the philosophical analysis of technical expressions such as 'reduction' in general. In order to establish a common ground, a taxonomy of notions in the vicinity of the concept of reduction is introduced, and the target of this book is located within this taxonomy. It is then argued that an appropriate characterization of the concept of reduction builds upon an explication, rather than an empirical investigation of alleged cases of reduction, or upon mere stipulation. Finally, the relevant terminology is introduced.

2.1 Reduction, Elimination, and Monism – A Taxonomy

How should we approach an appropriate characterization of the notion of reduction? This partly depends upon what we mean by 'reduction'. Among philosophers, it is common courtesy to acknowledge that 'reduction' may come with many different intended meanings. The concept of reduction, as conceived of here, is supposed to reconcile diversity and directionality with strong unity, without relying on elimination. This characterization may help to some extent. However, it employs notions such as *unification* and *elimination*, so that one may fear that it tries to illuminate the obscure with the opaque. To ensure that the introduction of the present book's target is not flawed by ambiguities in the use of 'reduction', 'elimination' and 'unification', or the distracting associations these terms may provoke, we should try to distinguish different "kinds" of reduction, or different conceptions thereof, thereby pointing to a taxonomy within which the target concept can be located. In passing, this will enable us to clarify the conservatism/eliminativism distinction and shed light on the connection between reductionist commitments and scientific realism.

R. van Riel, *The Concept of Reduction*, Philosophical Studies Series 121,
DOI 10.1007/978-3-319-04162-9_2,

2.1.1 Reductionism and Monism

Reductionism, as conceived of here, is a specific form of monism. Reductive physicalists are monists, and some radical idealists are monists as well. Monism, as conceived of here, is the doctrine that there is one kind of entities and properties *only*. How does the monist deal with the fact that there are different sciences, each of which purports to be concerned with reality, and each of which comes with its own models, its own experimental techniques, and its own conceptual schemes? Adopting *reductionism* is one of the options the monist has in response to this question: A monist may be a reductionist in the sense that she believes that the objects postulated by at least some of the different sciences exist, and that these objects are all of the same kind, e.g. all physical, or mental in nature. In this sense, a reductionist, as envisaged here, is a *realist* about the objects or properties postulated by reduced sciences. What other options does the monist have? Obviously, the monist may be an anti-realist about the objects of reduced sciences. To illustrate this point, let us pick an example and focus on the monist's stance towards *biology* only, assuming that her monism is not of a form of "biologism". How does the monist deal with the fact that there is a science of biology, on her view a non-fundamental science? The monist will take two aspects of the existence of a science of biology to be particularly pressing: first, its relation to reality, and second, its pragmatic or epistemic value. We have already distinguished two options the monist has regarding the metaphysics of biology. The monist may buy into the metaphysical commitments that go together with accepting biology, and, hence, adopt a metaphysically realist stance, or she may deny that biological properties or kinds exist, and, hence, adopt an anti-realist stance towards biology. Reduction, as conceived of here, reconciles diversity and directionality with unity without relying on elimination, that is: it reconciles diversity, directionality and *a realist interpretation of the reduced level* with unity (among objects at the reduced and the reducing level).[1] The eliminativist, in the sense of 'elimination' employed in the slogan, opposes the realist interpretation of the reduced level. This use of 'eliminativism' has to be distinguished from another use of the term.

2.1.2 Two Forms of Eliminativism

Both, a realist as well as an anti-realist interpretation of biology are compatible with two rival *pragmatic* assumptions about biology: the assumption that we should stick to biology, and the assumption that biology is a science we should dispense with. These views can be described in terms of the *eliminativism/conservatism* distinction as well. *Metaphysical* eliminativism can partially be cashed out in terms

[1]Here, I merely refer to metaphysical anti-realism. Semantic realism might be combinable with metaphysical anti-realism. This distinction will be mainly ignored here.

of ontological commitment; *pragmatic* or *epistemic* eliminativism can partially be characterized in terms of whether or not a science (or, more broadly, a scientific discourse) is pragmatically appropriate:

(Metaphysical Eliminativism)	If metaphysical eliminativism about F-s is correct, then F-s do not exist.
(Pragmatic Eliminativism)	If pragmatic eliminativism about *a science of* F-s is correct then this is a science we can dispense with.

Conservatism, the denial of eliminativism, comes in two forms, too, corresponding to metaphysical and pragmatic eliminativism. For our present concern, these partial characterizations will be sufficient. Employing these different conceptions of eliminativism and conservatism as well as the notion of monism, we can give a sufficiently precise characterization of the notion of reduction we are interested in here and distinguish it from other notions of reduction. To do so, let me from now on restrict the use of 'eliminativism' (and, similarly, 'conservatism') to its pragmatic or epistemic form. Instead of employing the *metaphysical eliminativism/conservatism* terminology, I will talk about *anti-realism/realism* (for a specific domain).

2.1.3 Four Types of Monism – The Taxonomy

Based on these minimal characterizations of the distinctions between anti-realism/realism and eliminativism/conservatism, we can introduce a taxonomy of four monistic positions, all of which have been associated with some stance towards *reductionism*. I will again refer to the case of biology:

(Conservative Anti-Realism)	Monism is true, and there are no biological properties or kinds, but nevertheless, biology should be conserved.
(Eliminative Anti-Realism)	Monism is true, and there are no biological properties or kinds, and therefore, biology should be replaced by a lower-level science.
(Conservative Realism)	Monism is true, and there are biological properties or kinds and therefore, biology should be conserved.
(Eliminative Realism)	Monism is true, and there are biological properties or kinds, but nevertheless, biology should be replaced by a lower level science.

Let me briefly comment on these versions of monism, drawing distinctions captured in Table 2.1.

Eliminative Realism seems to underlie unificationist projects that are not committed to the assumption that high-level sciences are metaphysically flawed. If a unified science seems desirable on independent grounds, this may provide reasons for

Table 2.1 Monisms and reductionisms

Monisms and reductionisms	Anti-realism	Realism
Eliminativism	Monism-1: Eliminativism/replacementism	Monism-2: Unificationism
Conservatism	Monism-3: Epistemic/pragmatic non-reductivism	Monism-4: Conservative reductionism

abandoning metaphysically or semantically appropriate high-level special sciences.[2] This amounts to a reductionist form of what can be called *unificationism*. It will be argued later that in a sense *Eliminative Realism* can be combined with *Conservative Realism* – if we assume (as we should) that the normative components in these claims are to be qualified: In some respects, it is worth "conserving" (using) a reduced theory, in some others, reduced theories could be "eliminated" (it would be desirable if we could ignore them). There neither is conservatism nor eliminativsm *simpliciter* (see Sect. 9.2).

Versions of *Conservative Anti-Realism*, according to which high-level sciences are not true, but, nevertheless, play an important role and have to be conserved, is usually justified by pragmatic considerations (Van Gulick 1992, 2001) or by the somewhat similar assumption that idealization plays a crucial role in the construction of high-level sciences (Sklar 2003). Even though theories of high-level sciences are not literally true or metaphysically adequate, they nevertheless yield a useful picture of reality, a picture that is maybe even more useful for some domains of inquiry than any picture lower-level sciences could possibly yield. A theory of good governance describes reality using a conceptual scheme we can easily handle, which is, by and large, appropriate, even though the propositions expressed by sentences in a theory of good governance are not, and cannot be, literally true of the matter at hand. Being confronted with a true description of the relevant situation, however, we would be utterly lost. Therefore, we should stick to the pragmatically appropriate high-level science. This stance can be regarded as one form of non-reductive monism, namely, a pragmatic or epistemic version of this doctrine.

[2]Maybe in the early years of positivist behaviorism, we can find a position according to which at least some high-level sciences are possibly true though eliminable. If we take positivist behaviorism to state that (i) psychological terms are to be *analyzed* in behavioral terms, then positivist behaviorism is committed to the claim that there are psychological kinds (because there are behavioral kinds represented by psychological concepts). So, psychology is a science that is at least possibly true. If we assume that these positivists claim that (ii) psychology should, in principle, be replaced by a science that is formulated in the language of physics, because we can hope for an analysis which bridges the gaps between scientific levels, then we have a position that instantiates *Eliminative Realism*. Truth or appropriateness of psychology is combined with in-principle-elimination. To mention just two authors who could be understood in a structurally similar way: Carnap, assuming that we should seek at *translation* of all scientific statements (Carnap 1934, 32), and Neurath, who argues that we can understand utterances concerning, for example, empathy as being translatable into physical language (Neurath 1959, 298).

In contrast, (Eliminative Anti-Realism) is justified by reference to the alleged fact that (some) higher level sciences are so utterly misleading that they have to be replaced by lower level sciences. Those who subscribe to this kind of eliminativism normally argue that *merely some* higher level sciences should be eliminated, that is: subscribing to a version of (Eliminative Anti-Realism), no actual philosopher commits herself to the strong claim that *all* high-level sciences are equally defective. Trying to replace psychology by neuroscience makes sense only if neuroscience is regarded as fundamentally different from psychology, even though it is a higher-level science (it is not physics). Therefore, those who subscribe to some sort of *partial eliminativism*, like the Churchlands (Patricia Churchland 1986; Paul Churchland 1981, 1985),[3] should be regarded as subscribing to a weak or domain-restricted version of *Eliminative Anti-Realism*. This is classical eliminativism, a doctrine that usually cashes out the notion of reduction in terms of theory-replacement. To add another '-ism' to the world of labels for philosophical doctrines (and a particularly ugly one), we could call this view 'replacementism'.

I think that *Conservative Realism* is self-explaining: If the high-level science is metaphysically appropriate and, ideally, true, then why shouldn't we accept it as appropriate? The more challenging question in the context of both versions of realism is this: What is the relation between the higher-level sciences and the lower-level sciences if both are true and the fundamental level's entities and properties do all the work? Answers to this question range from ontological theories of supervenience and realization to theories of explanatory relations between theories. Note that, given the qualification of *Eliminative Anti-Realism* given above, you can be an eliminativist for some levels, and subscribe to *Conservative Realism* for others (and you can combine even more positions for different levels).[4] Conservative monistic realism is conservative reductionism.

In these brief characterizations, the term 'realism' has been employed in order to get rid of an ambiguity in 'eliminativism'. It should be noted that monism, and, hence, realism is metaphysically neutral; as already suggested, monism can be

---

[3]The Churchlands figure among the most prominent defenders of this sort of eliminative materialism. Eliminative materialism in this sense states that

> [...] our common-sense conception of psychological phenomena constitutes a radically false theory, a theory so fundamentally defective that both the principles and the ontology of that theory will eventually be displaced, rather than smoothly reduced, by completed neuroscience. (P. Churchland 1981: 67)

It is an interesting feature of this formulation that eliminative materialism is contrasted with reduction. The pre-theoretical notion of reduction presupposes that the reduced theory is true or appropriate, rather than false and totally misguided.

[4]Note secondly that for some philosophers it might be difficult to judge whether or not they belong to the camp of *Conservative Realism* or *Conservative Anti-Realism*: If you take truth to be described in terms of pragmatism, then, under certain interpretations, the distinction breaks down.

embraced by idealists as well as by naturalists. Similarly, the notion of reduction is, in a sense, metaphysically neutral, though particular forms of reductionism are not.

2.1.4 Reduction and Metaphysical Neutrality

To see that reduction is, in a sense, metaphysically neutral, consider the following questions: Why not define reduction in terms of physicalism? After all, as Melnyk put it, 'doctrines of physicalism [. . .] should be seen as competing responses to the many sciences problem' (Melnyk 2003, 2). Isn't physicalism, or naturalism, the form of monism reductionists adopt? And can an appropriate explication of reduction ignore the question of what physicalism consists in? After all, it is physicalists who prominently employ this term! It is common to qualify certain things as being physical. There are *physical* events, *physical* properties, *physical* kinds, *physical* laws, *physical* objects, and the *physical* level of a hierarchy of sciences. Accordingly, I will talk about *physicalism*, about *physical objects* and about *ultimate physics*. We have an idea of how to apply them, but we may not be able to give a definition. That is all we need for the moment – that there is a coherent use of these expressions we can rely on.[5] 'Physicalism', 'naturalism', and 'materialism' (but not its variants – there is no materialist level of a hierarchy of science) will be used interchangeably. I shall elaborate on related notions of being physical in Sect. 9.1.

Now, what is the relation between the concept of reduction, and doctrines of physicalism or naturalism? An appropriate definition of the notion of reduction could be given even if, in principle, we were unable to say anything about what physicalism actually states. Reductionism does not presuppose physicalism. This is due to the fact that the concept of *monism* is, in a sense, metaphysically neutral. Neither the notion of naturalism, nor the notion of idealism is built into the notion of monism. Since the concept of monism is metaphysically neutral, the concept of reduction is, too. That is: The reductive idealist and the reductive physicalist agree that there are relations of reduction in the actual world. They do not agree on the *direction* of reduction in the actual world. After all, an idealist who believes that everything reduces to the mental may employ the term 'reduction' in the way the physicalist employs the term! They do not disagree about the *notion* of reduction, but rather about which reduction relations actually hold, about how our world is shaped metaphysically, or what is more fundamental – the mental or the physical. Compare

[5]Definitions of being physical normally come in ontological terms, that is: in terms of what sort of objects and kinds are physical *simpliciter* and, sometimes, the physical is defined as the non-mental (Spurrett and Papineau 1999; Papineau 2002, ch. 1.10; Crook and Gillet 2001). Sometimes, philosophers defined the physical via reference to physics (Papineau 1993; Hellman and Thompson 1975, 553 ff.).

this dispute to the following dispute: Assume that two persons argue about whether or not the Eiffel-tower is taller than the Tower of London. They do not disagree about the meaning of '_is taller than_', but rather about what actually is taller – the Tower of London or the Eiffel Tower.

The notion of reduction is independent of the notion of physicalism at least in this respect: It is conceptually possible that (a version of) reductionism is true, and that physicalism is false. Possession of any richer concept of reduction, that combines metaphysically neutral reduction with physicalism, requires possession of the topic-neutral concept. So, we should not define our general notion of reduction in terms of physicalism. The notion covers a specific dependence relation that is neutral with respect to what actually forms the fundamental stuff. Physicalism is a doctrine about what the fundamental stuff is. Thus, these two issues have to be kept separate. The concept of reduction is *not* metaphysically neutral in the sense that reductionism goes together with metaphysical commitment; *realism* suits the reductionist as conceived of here. So, what is the precise connection between scientific realism and reductionism?

2.1.5 Reductionism and Realism

It should be obvious that questions about reduction occur only, or become at least seriously more pressing if we assume that the aim of science is to provide us with true propositions about what actually is out there. The reductionist's rationale for presupposing this picture is that if sciences were necessarily concerned with mind-dependent entities, or with constructions, which necessarily miss the way the world actually is, then questions regarding reduction would seem rather unimportant. That the best theory of the actual world might be dualistic, or that it might presuppose a mentalist monism, is threatening to the physicalist only if this theory's furniture matches the furniture of the world we actually live in. Otherwise, the question of dualism would merely be a question about how we do or are bound to conceive of the world, rather than a question about how the world is. So, to give reductionism, and, hence, the concept of reduction an interpretation according to which issues of reductionism and anti-reductionism are obviously pressing, I will presuppose that reductionists are committed to scientific realism (for the relevant domains).[6]

[6]Note that the generality requirement underlying the debate between anti-realists and realists is not of much interest for us. Some sciences might be such that they aim at truths (to use van Fraassens idea, see below), whereas others do not. Thus, a weakened form of the realistic picture sketched above is to a certain extent compatible with modest forms of what is often labeled 'anti-realism'. Van Frassen's idea of constructive-anti-realism, for example, is characterized by its taking theories to be understood in a literal sense (what I referred to as 'semantic realism'). If we grant in addition that at least *some* theories are true or at least aim at truth without assuming that it is *the* aim of science in general "to give us, in its theories, a literally true story of what the world is like" (van Fraassen 1980, 8), then for these theories questions regarding reduction are equally pressing. The point is this: Questions regarding reduction, as these questions are understood here and will

Now, the different positions sketched above raise different conceptual issues. This book deals with *monistic realism regarding high-level science*. For the monist, the idea of reconciling diversity and directionality with strong unity without relying on (metaphysical) elimination, or anti-realism, is particularly pressing. This leads to an important distinction between issues related to *reductionism* on the one hand, and *eliminativism*, or *replacementism* on the other.

2.1.6 Two Philosophical Challenges: Reduction and Replacement

If we take reductionism to consist in the position that high-level sciences are (merely) to be replaced, then the question of how the *concept* of reduction should be characterized is not philosophically challenging at all. If one theory is replaced by another theory then one theory is abandoned, and the other is used instead. That's it. Interesting empirical questions arise, for example, when we try to find out under which conditions scientists actually replace theories, or when we are interested in the *epistemology* of theory-conservation and elimination. In the conservative case, questions may arise as to *pragmatic or representational issues*, which are concerned with a clarification of the idea that a false theory can serve a scientific purpose. So, we have to distinguish between two different issues: Even though the notion of theory-replacement is as clear as a notion can be, it is not so clear under which conditions scientists *believe to be (and are) justified in changing their theories*. Answering these epistemological and procedural questions, other questions concerning inter-theory relations turn out to be interesting, like the question of *how false scientific representation can serve a scientific purpose* or of *how a false theory can be similar to a correct theory (under certain conditions),* or, more generally, *what it is for non-equivalent theories to resemble each other.* Investigating such questions is what is done under the heading 'reduction' in the sense of 'theory replacement'. However, this has got nothing to do with the *concept*

---

be developed below, are interesting only in the context of allegedly *true* theories. If people were able to give a true science of everything, no anti-realist would dismiss such a theory on the grounds that it is to be conceived of as being true. In this case, the question of how, intuitively speaking, higher-level theories relate to this fundamental science, is of much interest. If higher-level theories turned out to be accidentally true or appropriate (in the light of the fundamental theory) everybody, the realist as well as the anti-realist, would still have the problem of reduction. Nevertheless, even in this picture, a realistic understanding of science enters the game, even though it is a realistic picture of possible science, rather than of actual science.

Now, finally, even if you dismiss realism, you might still believe that the truth-predicate as well as 'exists' and similar expressions are interesting from an instrumentalist perspective. Then, again, you might be interested in a theory of reduction from an instrumentalist perspective, and just interpret 'truth' or 'exists' the way you prefer. Thereby, you will arrive at a structurally similar, anti-realist interpretation of reduction that still should play an explanatory role within the chosen framework.

of replacement – it has got to do with concepts of other relations connecting actual (or possible) pairs of theories, containing one element that has been or should be or should have been replaced by the other. The *concept* of reduction as replacement is not philosophically interesting, although *reductions* as replacements give rise to a number of philosophically challenging issues.

In contrast, consider positions of the type of monistic realism. For example, such positions give rise to the metaphysical question of what a scientific level is supposed to be, if sciences at different levels are true even though monism holds. This notion will be the target of the investigation to follow. It is a notion of reduction according to which the reduction of one science (or kind, or phenomenon, or term, or fragment of discourse) to another implies that for the relata of the reduction relation, there are no two distinct levels of reality, or different sets of kinds and substances. Schematically: *a* reduces to *b* only if *a* and *b* are (ontologically speaking) homogeneous (and not trivially so, due to being empty). That is: If '*a*' and '*b*' are substituted by expressions referring to theories, then these theories' terms do refer, but do not refer to kinds at different levels, and if they are substituted by kind-terms, then the corresponding instance of '$a = b$' is true. Thus, if we were able to model this notion, we would be in the position to give an answer to the question of *how it is possible that there are true, ontologically committal sciences (or, more liberally: appropriate descriptions and terms) that are seemingly unconnected, or are situated at different levels, although they are compatible with strict monism.*

The notion of identity-based reduction this book focuses on thus fits monistic realism: *The concept of reduction is supposed to reconcile diversity and directionality (mirrored by the idea of levels) with strong unity (mirrored by the idea of monism), without relying on (metaphysical) elimination (that is the idea of realism).* But isn't this target, or this description of the target obviously ill conceived? Apparently, it does not do justice to the fact that in the philosophy of science, reduction is commonly regarded as a relation holding primarily between theories.

2.1.7 Theory Reduction and Ontological Reduction

Reduction in the philosophy of science is usually regarded as being a relation instantiated by pairs of theories, or scientific models, rather than by pairs of properties, event-types or states of affairs. It is thus time to distinguish between two versions of reduction – *theory* reduction and *ontological* reduction:

(Theory Reduction) The predicate '_ reduces to _' expresses a relation of theory reduction only if its arguments in true sentences refer to things such as theories, or scientific models, or fragments of theories.

(Ontological Reduction) The predicate '_ reduces to _' expresses a relation of ontological reduction only if its arguments in true sentences refer to things such as kinds, types, properties, events, substances, or individuals.

Obviously, the notion of reduction that gives rise to the puzzle sketched in the introduction rests on a version of ontological reduction, whereas notions of theory reduction are more holistic in spirit and can account for diversity in a straightforward way, namely, by referring to differences between the reduced and the reducing theory. For the moment, we will ignore this issue. I will treat the notion of reduction discussed here as distinct from a notion of theory reduction. Whether or not it collapses into a notion of theory reduction, or whether or not it is totally alien to the notion of reduction pertinent in the philosophy of science will be discussed later. Both questions will be answered to the negative in Chaps. 6, 7, and 8; in fact, it will turn out that "ontological" reduction is more fundamental than theory reduction.

When it comes to characterizing notions of reduction, suggestions are made by philosophers taking methods of the philosophy of language tradition seriously (e.g. Kim 1992; Lewis 1972), whereas philosophers of science, trained in formalisms, frequently conceive of the problem as a problem to be solved in formal terms (Moulines 1984; Balzer 1984; Sneed 1971). Some adopt a pragmatist stance towards questions about some epistemic or other value of high-level sciences (Van Gulick 1992, 2001; Sklar 2003), whereas others feel pressed to back up notions of scientific levels and their interrelations by telling an ontological story (e.g. Fodor 1974; Esfeld and Sachse 2007; Bechtel and Richardson 1993; Bechtel 1994; Craver 2005). Such differences in the format of the characterizations pose difficulties, but they do not necessarily result in fundamental problems. In contrast, there is a more problematic issue, not concerning the way of designing the characterization (putting it into structuralist or pragmatist terms, or giving a formal or a non-formal definiens), but rather concerning the method of approaching the characterization. So, back to our initial question: How should we approach an appropriate characterization of the notion of reduction?

2.2 Ways of Defining Reduction

To characterize different version of definitions of reduction, we need an idea of what is meant here by 'definition'. Assume that any candidate for a full-blown characterization comes in the following form:

x reduces to y iff xRy [. . .].

This is what I will call a *definition*.[7] For what follows, every bi-conditional connecting open sentences will be regarded as a definition. Let us talk about *kinds*

[7]The way the notion of a definition is introduced here will serve our present purpose. It is not supposed to generalize to other contexts. I will also be tolerant about answers to the question of *what* can be defined. Let us take predicates as a starting point. We can then arrive at derivative uses of the term 'definition'. A concept is defined iff some predicate expressing the concept is defined. A relation or property is defined iff the predicate picking out the relation or the property is defined (this might result in different true definitions of one and the same property, namely, when two

of definitions in order to distinguish between different interpretations of how a definition should be approached. There are basically three kinds of definitions philosophers have offered for the concept of reduction: *stipulating* a meaning for 'reduction', tying a characterization of the notion to an investigation of *alleged cases of reductions*, and, thus, taking an empirical dimension into account, and, thirdly and only implicitly, by way of *explication*. An explication can basically be conceived of as a procedure that tries to illuminate a term's pre-theoretic use by (i) reflecting upon criteria of adequacy this use imposes on a definition, without relying on the assumption that we can come up with a full-blown conceptual analysis of that notion, and (ii) by *sharpening* that concept. This section argues that stipulations as well as empirically inspired investigations of actual theory-successions do not form interesting candidates for an appropriate definition of the concept of reduction. Rather, we should try to *explicate* the concept of reduction.

2.2.1 Definitions as Stipulations

In some contexts the meaning of the term 'reduction' is treated as being fixed by an explicit definition or stipulation. Philosophers of science sometimes talk that way when they take what has come to be known as the Nagel-model of reduction as being fixed (stipulated) by Nagel's own definition of 'reduction' (utterly briefly): reduction is derivability plus bridge-laws, whose job is to connect the terms of the reducing science('s laws) to the terms of the reduced science('s laws), if the vocabulary of these sciences is heterogeneous (Nagel 1961, chapter 11). This has a puzzling effect: For example, Richardson in his (1979, 1982) seems to *presuppose* the Nagel model, rather than to take it as a *candidate* for a definition of the notion. Richardson argues that the Nagel-model is fulfilled by theory-derivation on the basis of bridge-laws that take the form of conditionals (something observed by Nagel himself (1961, 355, fn. 5)). Does this, without further ado, show that *reductions* can be based on bridge laws that take the form of mere conditionals? It does not; it shows that *Nagel's characterization* can be satisfied in this specific way. Whether or not conditions on *reduction* can, and, thus, whether or not the definition captures reduction, is to be decided on independent grounds. If a given pair of, in this case, theories does, intuitively, not fall under the concept, but does fulfill the definiens offered for this concept then we have reason to deny that the definition actually covers the concept. Being an instance of some model or definition of a concept C is *not* to be treated as being the make or break for being an instance of C. On some occasions, Kim (1993, 150 & 248) and Fodor (1981, 150) seem to follow this way of talking about (Nagelian) reduction. Recently, Butterfield has argued in a similar spirit that reduction and emergence turn out to be compatible, on

predicates with different meanings pick out the same property). A general term is defined iff the predicate that is build by (the copula and) the general term is defined.

a Nagelian interpretation of reduction (Butterfield 2011a, b). Interestingly, some versions of *reductionism* have been criticized based on a similar strategy, i.e. (implicitly) taking candidate characterizations to be stipulations. This point has been brought out by Gene Witmer in his intriguing discussion of Steven Horst's book[8] *Beyond Reduction: Philosophy of Mind and Post-Reductionist Philosophy of Science* (Horst 2007; Witmer 2008). Horst argues that reductionism is mistaken *because* reductions in the sense of the Nagel-model are rare (Horst 2007, 49). But what if the Nagel-model did not adequately capture the relevant relation? In this spirit, Witmer objects that failure of the Nagel-model does not translate into a failure of reductionism – he suggests that it 'should still be open to us to say that A is reductively explained by B even though a derivation is impossible' (Witmer 2008).

What has this got to do with stipulation? You cannot make a mistake *stipulating* that reduction is derivability plus bridge-laws. But you can make a mistake *proposing* that reduction is derivability plus bridge-laws. For stipulations, it is just pointless to ask questions about whether or not the definition is appropriate. If the meaning of 'reduction' were to be fixed by stipulation, then there would not be a substantive issue here. This sort of "approach" (which is barely worth this name in the case of reduction) will not loom large in the rest of this book. It just misses the target. As a desideratum, we get that a definition should be approached in a way that allows for mistakes – we want a serious discussion about whether or not a candidate definition is appropriate.

An alternative to stipulation is tied to the idea that part of the job of philosophers in the philosophy of science is to *rationally reconstruct* science, or to do an exercise in *meta-science*: reflect upon actual cases of theory-successions to shape a model of reduction.

2.2.2 Empirically Inspired Definitions

Let us begin with a very radical point of view: that reduction is to be defined in accordance with a previously fixed set of examples. This idea appeared in one of the first formulations of how to approach the notion of reduction in the literature. Kemeny and Oppenheim (1956) write:

> The label 'reduction' has been applied to a certain type of progress in science. As this process has been the subject of much philosophical controversy, it is the task of the philosopher of science to give a rational reconstruction of the essential features of reduction. We will discuss the basic features of this process informally, we will review two previous attempts to make the concept precise and we will offer certain improvements which we hope will *bring the philosophical characterization of reduction closer to what actually happens in science*. (Kemeny and Oppenheim 1956, 6, my emphasis)

Kemeny and Oppenheim clearly state that they believe the concept of reduction to be defined in a way that gives an insight into actual scientific processes – it is

[8] I am grateful to an anonymous reviewer who brought this book to my attention.

a technical term the definition of which has to be judged by how well it fits a set of previously fixed examples. In this context, the term 'reduction' is reserved for *whatever the relevant relation might be which holds between different theories or sciences, which stand in a succession-relation to each other (and share, maybe, some additional features)*. Balzer (1984, 331) explicitly reflects on this point without deriving a definite conclusion. According to him, we should conceive of approaches to the notion of reduction as consisting in reflections on actual examples, which he describes as *paradigmatic* cases from which a tentative definition is derived, which is then applied to alternative cases. If the definition does not cover a case we believed to be covered, we have to decide whether or not the case is to be counted as relevant. So, Balzer, Kemeny and Oppenheim seem to share the intuition that any definition of the notion of reduction is to be judged by its appropriateness with respect to a certain class of examples. This can be subsumed under the label of *'extension first'* approaches to the definition of reduction. Even though the techniques employed are often formal (and the theory-relations so defined are formal), these models' touchstone is the adequacy for a given set of actual theories. Finding out about whether or not they match these set's elements is an empirical matter.

A cognate of this approach is a *mixtum compositum* of experimental philosophy and rational reconstruction. What John Bickle labels 'meta-science' (Bickle 2008) can roughly be characterized as follows: In order to understand what reduction is, we have to pay attention to how scientists use the term, i.e. what they claim to be a reduction. The philosopher's job is to pay attention to how scientists use the term, and to reconstruct the cases so described by scientists, in a way that is similar to rational reconstruction as described above.

Note firstly that, as argued in Sect. 2.1.6, no *concept* of theory-change is philosophically challenging (although the epistemology or the logical relations between the *relata* of theory-changes are challenging). Similarly, Bickle's project is not a conceptual one. The concept of reduction is, on this account, not challenging at all; the challenge lies in understanding the cases described by scientists as reductions. The main difference between classical rational reconstruction and meta-science seems to be this: In the classical case, it is the philosopher who decides which case of alleged reduction to include in the set of relevant cases. On Bickle's interpretation, it is the scientists' use of the term 'reduction' that defines this set.

Note secondly that results of this procedure will not be very satisfactory, if we aim at a definition of a notion that is already in use in specific explanatory contexts *in philosophy*. As long as there is no criterion by which we can judge whether or not the decision concerning the relevance of an example is appropriate, this sort of approach is at risk of involving an element of *ad hoc* decisions about what to count as belonging to the concept's extension. Bickle suggests a criterion; but one may wonder what the scientist's authority concerning the use of the term 'reduction' in philosophy is. The concept of reduction seems to play an important role in characterizing reductionist positions and in clarifying variants of non-reductive physicalism. The term 'reduction' thus has a use that is prior to any definition. It is hard to see how an investigation of actual cases of theory-successions (or similar cases of explanations in science) can, without further ado, do justice to this pre-theoretic use. As a desideratum, we thus get that our method should do justice to the pre-theoretic use of the term.

However, in a sense, Kemeny's, Oppenheim's, Balzer's and Bickle's dialectical remarks should be taken seriously. As we shall see in Chap. 7, attempts to include replacement in a definition of reduction can be regarded as being carried out in the spirit of this way of talking about reduction. The very fact that Kemeny, Oppenheim and others use the term 'reduction' in this way is sufficient to claim that in fact, there is a technical term 'reduction' in the sense just outlined. But this is not the only one.

2.2.3 *Definitions as Explications*

Sarkar contrasts attempts to model a certain relation we have access to independent of empirical investigations, with attempts to come up with accurate descriptions of a set of actual scientific developments (Sarkar 1992, 169; see also Wimsatt 1976). We just dismissed the second approach for the context at hand. But what is the alternative? Doesn't the fact that intuitively, this approach should be independent of empirical considerations, at first sight, make it look like good old conceptual analysis?[9]

Block seems to deny that conceptual analysis forms a promising candidate for approaching a definition of reduction, stating that there is "much less interest in analyzing technical philosophical concepts [than there is in analyzing some ordinary language concepts]. We should use whatever technical concepts do the jobs we want done." (Block 1997, 112) This rough characterization already hints at the main idea underlying the concept of an explication: There is a job a technical concept is supposed to do. This job has to be identified. An appropriate characterization should reflect an appropriate job-description. Intuitively, we thereby combine an element of stipulation with reference to previously fixed criteria of adequacy – the latter characterizing the job we want our concept to do. This makes room for mistakes in definitions. Consider the following statement, where Dizadji-Bahmani, Frigg, and Hartmann argue that

> … the syntactic view is unnecessary to get GNS [the *Generalized Nagel Schaffner Model*] off the ground […] Where first order logic is too weak, we can replace it with any formal system that is strong enough to do what we need it to do. (Dizadji-Bahmani et al. 2010, 403)

[9] Nickles (1973) seems to at least partly rely on the idea that we should reflect upon how the term 'reduction' is used in different contexts in order to come up with an appropriate characterization. His main point is that there are different uses of the term in philosophical and scientific contexts. The notion focused on here is closer to what he labels 'reduction_1' rather than to 'reduction_2', that captures, intuitively, some aspects of actual scientific change.

The authors seem to adopt a view according to which it is possible that one might be *mistaken* about a definition's appropriateness. The authors rescue what they take to be the core idea of Nagel's (and Schaffner's) model of reduction by abstracting away from *irrelevant* or *misguided* positivistic aspects of the definition. This is possible only if the definition is not a stipulation. Secondly, they describe the definition achieved by the suggested amendments as *Nagelian* (being in accordance with the Nagel-model) – it is *Nagelian* in the sense that it captures what Nagel aimed at. Accordingly, there is a core idea which can be *distinguished from* the definition and which is *prior to* the definition.[10]

This idea is similar to the idea of (early) *Carnapian explication* (Carnap 1950, 1988 (1956), §2). The idea of explication is captured by the following passage, where the *explicatum* is a newly introduced expression, which is introduced to replace the *explicandum* in a certain context:

> Generally speaking, it is not required that an explicatum have, as nearly as possible, the same meaning as the explicandum; it should, however, correspond to the explicandum in such a way that it can be used instead of the latter. (Carnap 1988 (1956), 8)

Carnap's conception of explication is not particularly precise. He illustrates the explication relation referring to the relation between the ordinary-language concept of heat and that of temperature (as a measurement-concept) and he introduces four criteria any appropriate explication should meet that can be summarized as follows: *similarity in use*, *simplicity of the definition of the explicatum*, *fruitfulness of the definition of the explicatum*, and *exactness of the explicatum* (1950, 7 ff.). Arriving at measurement concepts plays as key role in his conception of explication. His own suggestions for explications are, apart from his work on degrees of confirmation, often radically different from the heat-temperature example; he is concerned with the explication of ordinary language concepts, such as truth and necessity, by concepts defined for a formal language, that should be used in this formal language (the idea is *not* that they should replace the ordinary language notions in ordinary language contexts). We do not learn much about the typical deficiencies of concepts that should be explicated; Carnap just mentions that a concept that is worth explicating is usually vague (Carnap 1988 (1956), 7 – where 'vagueness' should probably not be understood in the linguist's sense of the term).

Quine (1960, § 53) takes up the Carnapian idea, arguing that explication should replace conceptual analysis as classically conceived – that is: as decomposition of meaning constituents (or something similar), which obeys criteria of synonymy, as he puts it (Quine 1960, 258) – in *any* context. Fortunately, his remarks on the relevant similarity relation between the explicandum and the explicatum are more precise than Carnap's. According to Quine, we should conceive of the relation between explicatum and explicandum in terms of *functional* similarity. Some functional features of the explicandum are to be preserved by the explicatum, whilst

---

[10]Similar ideas are pertinent in Endicott (1998), Waters (1990), and it is discussed in van Riel (2011).

others are to be neglected. Since these function-preservations are not to be achieved by complete meaning-conservations, an element of stipulation enters the game, as his discussion of the definition of the term ‘pair’ nicely illustrates (Quine 1960, §53). Quine suggests that the “ordinary” concept of a pair is just deficient – the metaphysical commitments that go together with the use of this concept are just untenable. Nevertheless, aspects of this concept are worth preserving. In successful explications of the concept of a pair, we get definitions that preserve the desired aspects and get rid of the misleading ones.[11] So, the upshot is this: If there is an expression which serves a certain important goal, but which poses serious philosophical problems, we should try to replace this term by another term (give this term another use), by introducing a definition which involves an element of stipulation, but which, nevertheless, is oriented towards the goal the primary use of the term determined. This orientation towards the same explanatory or descriptive goals is guaranteed by partial functional equivalence between explicandum and explicatum.[12]

In the present context, we are neither concerned with formal languages which are supposed to model natural languages (in certain respects), nor should we conceive of the task of defining the notion of reduction to be the task of replacing a not quite exact, or vague, or ontologically misleading expression by a term expressing an exact formal concept, which can be used instead of the previous expression in a certain context. To repeat: ‘Reduction’ is a technical term, and we just do not know how to define it, such that it can perform the job it is supposed to do. However, the procedure associated with explications seems promising when adopted for the case of ‘reduction’.

In order to ensure that our concept can perform the job it is supposed to do, we start with non-formal characterizations that define a set of criteria S (this corresponds, intuitively, to the concept’s *job-description*). We grasp these criteria reflecting on *meaning*, or reflecting on *use*. For example, the notion of reduction, as it is introduced in the literature, describes a directional, or a seemingly asymmetric relation. Whoever fully understands the pre-theoretic characterizations is in a position to know that if ‘*a* reduces to *b*’ expresses a truth, then ‘*b* does not reduce to *a*’ expresses a truth as well. Up to now, the project is similar to classical conceptual analysis. In a second step, however, we try to introduce a notion that satisfies a set of criteria $S^* \subseteq S$ (in the best case, $S^* = S$, if S does not comprise contradictory, irrelevant, or otherwise problematic criteria). The conceptual resources we use to match these criteria depend on our decisions and, occasionally, on reflections upon independent issues. The idea of reflection upon independent issues can be canvassed as follows: There might be a number of definitions which match the criteria equally well, but which differ in other important respects, especially with respect to how they conceive of *other* notions that enter the definition. Let me give an example. Assume that Nagelian or classical empiricist or syntactic reduction

[11] Say, by Kuratowski’s set-theoretical characterization according to which <a, b> = {{a}, {a,b}}.

[12] For a discussion of the notion of explication, see also Strawson (1963) and Hanna (1968).

(R_N) and structuralist or (more broadly) semantic reduction (R_S) are equivalent (in a sense that will be specified in a moment). In this case, they differ primarily with respect to how they conceive of *theories* or *scientific representation*, and, correspondingly, of derivation, but are *equivalent in other respects*. In this case, we have to take independent considerations into account in order to decide between these definitions. Here is an idea of how we can conceive of a relevant kind of equivalence between two models of reduction in a very simplified manner. Assume that for any structuralist theory, t, there is some (possible) syntactic theory, t^*, which stands in some relation R to this structuralist theory, such that tRt^*. Moreover, there is no other structuralist theory $t' \neq t$, such that $t'Rt^*$. Then, there is an interesting similarity between the notions of reduction defined in terms of these different kinds of theory iff for any reduction of a structuralist theory, there is a corresponding reduction of a syntactic theory, and vice versa. Intuitively, we can think of such a relation between syntactic and structuralist theories in terms of *expression*. Let the structuralist theories be the structured semantic values of syntactic theories. Note that this is not quite appropriate, because classical empiricist theories do not simply express structuralist theories (or structures). If we assume that there are enough (possible) syntactic theories, however, *expression* is a perfect dummy. Here is a semi-formal version of this idea of an equivalence between the two reduction relations R_N (Nagelian or syntactic or positivist) and R_S (structuralist), where 'x' and 'y' range over theories in the structuralist sense and 'x^*' and 'y^*' range over theories as syntactic entities (ignoring modalities):

$$\forall x\, \exists x^*\, (x^* \text{ expresses } x\ \&\ \neg\exists y\, (x^* \text{ expresses } y\ \&\ y \neq x))$$
$$\forall x, y, x^*, y^*\, ((x^* \text{ expresses } x\ \&\ y^* \text{ expresses } y) \rightarrow (xR_S y \leftrightarrow x^* R_N y^*)).$$

This equivalence is interesting, because it will give rise to similar claims concerning hierarchies of sciences, concerning monism and so forth. If we assume that the notion of reduction does not presuppose a certain notion of a theory, then which relation we should choose to define reduction, R_N or R_S, depends on independent criteria, presumably concerning the metaphysics of theories. It might turn out that theories are to be conceived of as structures rather than syntactic entities, such that the structuralist definition is more appropriate in this respect. In this sense, an explication may involve aspects of stipulation.

Nevertheless, on this interpretation, explication has still got a lot to do with conceptual analysis. Intuitively, it aims at a partially *illuminating* definition of the target-concept, just like conceptual analysis, classically conceived, aims at an illuminating or meaning-revealing definition.

Definitions can be correct without being illuminating. A definition is correct iff the truth conditions fixed by the *definiens* capture the range of (possible) applications of the predicate so defined. However, there are correct though uninteresting definitions. Let us assume that the first natural kind I thought about this morning was the kind *horse*. Then, we can define the predicate '_is a horse' as follows: *x is a horse* $\leftrightarrow_{def.}$ *x belongs to the first kind I thought about today (in the actual world)*. This is a correct though non-illuminating definition. Similarly, and, avoiding the use of indexicals, we can define the concept of water as follows: *x is water* $\leftrightarrow_{def.}$ *x is* H_2O. If water is identical to H_2O, then this definition fixes the truth-conditions of applications of the predicate '_is water'. If you assume that it suffices to fix truth-conditions for the actual world only, we could go a step further: Assume that being a renate (the property of having kidneys) is not being a cordate (the property of having a heart),[13] despite the fact that all renates are cordates, and *vice versa*. Then: *x is a renate* $\leftrightarrow_{def.}$ *x is a cordate* is a perfect definition. Compare the examples given so far to the following case: *x is a vixen* $\leftrightarrow_{def.}$ *x is a female fox*. Again, the truth conditions are fixed. But, in addition, we learn something about the meaning of the predicate '_is a vixen' (as opposed to merely learning something about the predicate's extension, intension, or the property signified by the predicate). Put differently: This is a definition based on intuitions concerning meaning rather than a definition which is true in virtue of coincidence or mere a posteriori metaphysical truth. We can describe this definition as *illuminating*. Note that there are two kinds of illuminating definitions, the first of which is instantiated by the vixen-definition. It can be described as being based on a decomposition of meaning: The meaning of the term 'vixen' is given in the *definiens*. However, there are definitions which are illuminating (in an interesting sense), but which do not build on decomposition of meaning, even though they are based on relations between the meanings of the terms employed in the *definiens* and the *definiendum*. Here is an example: *x is a natural number* $\leftrightarrow_{def.}$ *x is either odd or even & x is a positive number (or x is zero)*. This type of definition is based on reasoning about meaning, but we should not regard the *definiens* as consisting in the decomposition of the meaning of the predicate used in the *definiendum*. It is illuminating in the sense that it gives us some insights into the meaning-relations between different predicates, unlike the other examples, but it is not illuminating in the way the vixen-example is. Why is this distinction important? Even though meaning-decomposition might be a project not worth aiming at in the context of reduction, it might still be the case that a definition based on meaning considerations can illuminate some aspects of the notion as it is pre-theoretically, or informally characterized. Ideally, these characterizations yield the relevant *job-description*. By becoming clear about this description, we give a (partly) illuminating definition, which may involve stipulations. This approach does justice to the desiderata identified above: Candidate definitions can be mistaken, and in order to avoid mistakes, we should pay attention to the term's pre-defined use, namely, by paying attention to the job-description. But which job-description can be extracted from the informal use of the term 'reduction'?

[13]This example is taken from Quine (1986 (1970), 8 f.). Note that Quine is concerned with the question of *synonymy*. Here, we are concerned with *definitions*. Both questions are intimately connected, but they are distinct. For the present purpose, I slightly changed the example: Property designators (like 'being a renate') do not figure in the Quinean case. Talking about properties, it seems easier to construct the relevant example.

2.3 The Concept of Reduction as the Subject of Explication?

We are already familiar with a short description of the job 'reduction' is supposed to do: *The concept of reduction reconciles diversity and directionality with strong unity, without relying on elimination.* Let me briefly connect this slogan to informal characterizations of the term that can be found in the literature, and then turn to alleged paradigmatic cases of reduction. We thereby arrive at and motivate the relevant criteria of adequacy. Here are some of these informal or pre-theoretic characterizations, which will lead to criteria of adequacy an explication of the notion of reduction has to fulfill:

(Kim)	If Xs are reduced, or reducible, to Ys, there are no Xs over and above Ys. (cf. Kim 2006, 275 f., (given in a similar fashion by Smart 1959))
(Wimsatt)	Inter-level reductions are compositional. They localize, identify, and articulate mechanisms that explain upper level phenomena, relationships and entities. (Wimsatt 2006, 449)
(Chalmers)	[W]hen [in the context of a reductive explanation, *RvR*] we give an appropriate account of lower-level processes, an explanation of the higher-level phenomenon falls out. (Chalmers 1996, 42)[14]
(Nagel-1)	[In reductions, A] set of distinctive traits of some subject matter is assimilated to what is patently a set of quite dissimilar traits. (Nagel 1961, 339 f.)
(Nagel-2)	[The reduced] science deals with macroscopic phenomena, while the [reducing] science postulates a microscopic constitution for those macroscopic processes. (Nagel 1961, 340)
(Sarkar)	The reduced theory is explained by a reducing theory which is presumed to be more fundamental. (Sarkar 1992, 167)

These descriptions are often enriched by examples, which supposedly describe *paradigmatic cases* of reduction: (*human*) *pain* reduces to *C-fiber stimulation*, or *water* reduces to H_2O (the latter can be found in (Putnam 1975, 431) (Fodor 1981, 150) and (Kim 1992, 23)).[15] Talk about levels, part-whole relations and explanation suggests *directionality towards*, or *priority of* the reducing item. Talk about assimilation and not-existing "over and above" suggests unity as well as

[14]According to Chalmers, reductive explanation is explanation of a phenomenon in terms of a set of properties on which the property of being an instance of this phenomenon globally logically supervenes (Chalmers 1996, 48).

[15]Note that these cases are paradigmatic in a sense different from the sense in which Balzer's cases are paradigmatic. In Balzer's case, the paradigmatic cases are actual cases, and if it turns out that an actual case, which was used as a paradigmatic case (in Balzer's sense) does not match the definition, the definition is to be changed. In contrast, these paradigmatic cases can be used even if water is not H_2O.

non-elimination. This is also reflected in the *use* of the term 'reduction'. Let me give just one example to illustrate what I am thinking of:

(Menzies) It is widely held among philosophers that macrolevel causal relations are reducible to microlevel causal relations. (Menzies 1988, 551)

This statement embeds a claim concerning the relation between two different sets of causal relations. The term 'reduction' is *used to articulate a thesis about a reductive relation.* It is not a characterization of the use of the term 'reduction'. It should be clear that what underlies "what is widely held among philosophers" will, in this case, be some vague intuition concerning reduction, or a family of intuitions, rather than some clearly defined concept of reduction. Similar statements include 'mental properties reduce to physical properties', or 'wholes reduce to their parts'.[16]

So, what role does this set of descriptions, paradigmatic examples and uses of the term 'reduction' play in the reduction debate? The answer is simple: Descriptions and paradigmatic cases serve the purpose of giving a first, tentative idea of what a well-defined notion of reduction should roughly consist in. These descriptions express intuitions and preliminary ideas of how the notion of reduction is to be conceived of. Paradigmatic cases can be regarded as the (or a more or less flexible version of a) touchstone for any definition of reduction. Using the term in philosophical discourse that is not concerned with an introduction of the term imposes constraints on the job-description.

We have thus sketched one reason to assume that the third variant of developing a definition is appropriate in the present context. There are pre-theoretic characterizations that should *guide* the explicit definitions. These characterizations describe the job our concept of reduction is supposed to do. It is not a set of examples, which does so, but rather a number of criteria we already possess when we try to define the notion of reduction. The role of examples, or paradigmatic cases is this: Independent of whether or not water actually is H_2O – we just assume that it is. Then, this case serves as a good example of a reduction. No empirical investigation of this example will give us a deeper understanding of the concept of reduction. This stands in stark contrast to what Kemeny and Oppenheim had in mind. As suggested above, the message of approaches that are in the spirit of Kemeny and Oppenheim can be captured by labeling them *extension first-approaches.* In contrast, we aim at a definition based on criteria that are gained from intuitive characterizations of the concept, or its explanatory role in philosophy – hence, the approach can be characterized as a *criteria first-approach!* The notion of reduction is, on this latter interpretation, at least partly determined by criteria that are gained independently of reflection on (an alleged) extension. Whereas *extension first*-approaches suit the goal of a rational reconstruction of scientific endeavors, such as *reductions*, any illuminating explication of the philosophical *concept* of reduction will have to rely on a *criteria first* approach.

---

[16]These statements do not perfectly fit other examples of reduction-talk; I will later refer to these statements as 'generic' statements – we need a special apparatus to deal with these constructions (see Sect. 5.6.2).

There is a use of the term 'reduction' we have not commented upon yet that clearly matches the goal of extension first approaches: When philosophers speak of *asynchronous* reduction, they have in mind a specific kind of theory-change. This concept captures a temporal relation that involves theories (or stages of one theory) at different times. In contrast, one may tentatively describe the notion of reduction we are interested in here as *synchronous*, or, in a more Platonist spirit, as *atemporal*: It is a relation that does not require its relata to instantiate a certain temporal relation, such as the relation of *occurring later than.*

Our slogan contains the core idea that defines the concept's job (schematically):

(Job-Description): We need a definition that

(i) accounts for the directionality, such that it gives a definition according to which if *a* reduces to *b*, then necessarily, *b* does not reduce to *a*, and
(ii) accounts for the idea of unity (ideally in the sense of strong unity) without elimination.

In addition, it will be desirable that our definition enables us to

(iii) illuminate the paradigmatic cases;
(iv) explain the intuitive characterizations of the notion of reduction;
(v) account for related topics, such as reduction and physicalism, reduction and scientific unification, and similar issues; i.e. the explication should yield a fruitful notion by fulfilling the explanatory task associated with 'reduction'.

Let me add this proviso: In contexts of explication, one may feel free to drop one criterion in favor of another. One case in point is this: As already mentioned, in the philosophy of science, it is common to drop an assumption imposed by the paradigmatic cases, namely, that reduction should connect *kinds* or *phenomena* or *events* (in any primary sense of the term 'reduction'). Approaches to reduction that take theories to be the primary *relata* of the reduction relation follow this idea. This is usually justified by arguing that the idea of "ontological" reduction is just confused. We will turn back to this assumption later on. Let me close this introduction with a few terminological remarks.

2.4 Terminology

The most important terminological distinction for the discussion to come can briefly be described as the distinction between *ontological* talk and *representational* talk (for the application of a similar distinction to the reduction-debate, see Van Gulick 2001). Here is a bunch of ontological items philosophers in the philosophy

Table 2.2 Ontological furniture

Ontological furniture	
Abstract objects	*Concrete objects*
(not located in space or time)	*(located in space and time)*
Kinds	*Particularized properties*
Relations	*Particularized events*
Properties	*Tokens*
Types	*Compositions*
Sets	*Sums (of concrete objects)*
Events	*Wholes*
	Parts
	Individuals

of science and in the philosophy of mind often refer to: properties, relations, kinds, types, events, sets, sums, compositions, wholes, parts, tokens, particularized properties, particularized events, and individuals. Let me propose a list that is structured with respect to the ontological location of its items. The list is Platonist in spirit, even though, with small changes, it allows for both, nominalist as well as constructivist readings. Table 2.2 gives an overview on our ontological furniture (I will comment on the representational furniture in a moment).

These categories are not mutually exclusive. Any individual horse, for example, is a token of the natural kind horse; we might conceive of compositions as a certain kind of wholes, for example as structured wholes, and of wholes and compositions as individuals; we could conceive of events as properties, and of particularized events as particularized properties.[17] For what follows, 'property', 'type' and 'kind' will be used interchangeably. The term 'relation' refers to what can also be referred to by 'relational property', 'relational type' and 'relational kind', or, briefly, by 'kind', 'property' and 'type' (thus, there are n-ary kinds, types and properties with $n > 1$ as well as with $n = 1$). If you prefer a nominalist ontology, take properties to be sums of individuals (and replace the idea of properties not being located in space and time by some other idea that serves your purpose), if you are more liberal, take them to be sets of entities, if you are even more liberal, take them to be functions from worlds to extensions. For the discussion to come, it will be useful to think of properties in one of the intensional senses or, for most of the properties the bearers of which are individuals occupying a position in space and time, in functional terms.

The purpose of this list is primarily to indicate that if one of its items occurs, we have a case of ontological talk. Secondly, we can use this list to introduce relations instantiated by pairs of items on this list: Particularized properties and

[17]Lewis (1986) treats events as properties (Lewis interprets them as properties of space-time regions), Kim interprets them as property-exemplifications which involve a property, a substance and a time (Kim 1976). Davidson famously defended the idea that events are particulars (Davidson 1969, 1970). Here, I shall speak of events as follows: there are event-*types* and there are *instances* of event-types, namely, concrete particularized events. Thus, I shall assume that at least in principle, talk about kinds of events makes sense. Further differences will not matter for our present concern.

Table 2.3 Representational furniture

Representational furniture	
Conceptual level	*Vehicles*
Contents	*Words*
Meanings	*Sentences*
Propositions	*Theories*
Concepts	*Mental representations*
Theories	
Events	

particularized events *instantiate* properties or events. Parts *constitute* complex wholes and compositions, and different things *form* a sum of these things. In addition, we need a convention concerning ways of making these things available in discourse: We talk about such entities using terms, or, sometimes, in the case of properties, using predicates. I will say that terms and predicates *pick out* the entities we talk about using these linguistic devices. Whereas terms *designate* or *refer to* the entity they pick out, predicates *signify* the properties they pick out. Possible differences between *referring terms* and *descriptions* that *designate* objects will be neglected here.

In contrast, terms and predicates *express* their conceptual content, if any, which brings us to our second list containing the representational furniture, given in Table 2.3. It involves what is expressed as well as what expresses, that is, entities such as concepts on the one hand, and words and sentences on the other.

Note that these lists are not complete. They form a device used to indicate representational talk as being distinct from ontological talk. Obviously, we can talk about representational entities in an ontological way, for example, when we use a term to refer to an expression's meaning, or to a word. That is: The representational entities can become part of the ontological furniture of a discourse, namely, a discourse that is concerned with representational issues. Meta-language is a paradigmatic example of this kind of discourse.

Relations of *picking out* can be regarded as being concerned with a language's ontology, whilst the relation of *expressing* is here understood as being concerned with a language's conceptual realm, or with a language's meanings. I shall use the term 'concept' in constructions which are used to refer to meanings (like 'the concept of a horse'). Intuitively, one can describe the difference between a language's ontology and its conceptual realm in terms of the objects, facts, and properties a language is *about* on the one hand and the representational structure in virtue of which it is about these entities and properties on the other. However, whether or not some feature belongs to the language's ontology or to its representational level is dependent on the discourse, not (primarily) on what the entity is. Note that this is not to say that any expression has a meaning, or that if an expression picks out an object or a property, it has a meaning beyond what it picks out. Directly referring terms lack meaning (in the sense of 'conceptual content').

The term 'theory' occurs twice in Table 2.3. This is neither to be regarded as a stipulation, nor as a substantial thesis about what theories are. Rather, it mimics two different ways of talking about theories. In the second sense (as linguistic entities or representational vehicles), theories can be regarded as stets of sentences, that is: sets of linguistic entities. Strictly speaking, they are therefore not subject to operations such as derivation, and they do not stand in relations like the implication relation to each other, relations which hold in the primary sense of the terms 'derivation' and 'implication' between linguistic entities only (no set is a linguistic entity). Nevertheless, I will, in accordance with the literature, say that sets of sentences *imply* sentences or other sets of linguistic entities.[18] Moreover, I will ignore the aspect of theories as *developing* entities. Thus, I idealize, using a static notion of a theory.

Now, we are in a position to generate a number of operators, which will enable us to distinguish between ontological aspects and conceptual aspects of a certain linguistic item. Talking about properties, I will use the singular term forming operator 'the property of being _', which works on general terms (similarly, I will I use 'the _ relation' for general terms which, if they are used to form a predicate in the standard way, can form an n-place predicate with $n > 1$).[19] The result of an application of this operator to a general term is a property designator that refers to the property that is picked out by the general term (or is signified by the predicate that is composed of the copula (if any) and this general term). Sometimes, and when appropriate, I will use the suffix '-ness' to form a property-designator. 'Red*ness*' refers to the property of being red, the property that is picked out by 'red' and signified by '_is red'. Note that sometimes, it will be suitable to treat terms like 'water' and 'H_2O' as referential expressions designating kinds; for example, when we say that *water* reduces to H_2O, this should in at least some contexts be interpreted as stating that the kind water reduces to the kind H_2O. (Similarly for 'pain' and 'C-fiber stimulation'.) Thus, 'water' is, at least on some occasions, in the same boat as 'the property of being water'; whether or not it is will become apparent from context.

Similarly, I will refer to conceptual contents or meanings using results of an application of the operators 'the concept of a _', in the case of sortals, and 'the concept of _' in the case of non-sortals. These operators also work on general terms. The referents of such terms are the conceptual contents expressed by the embedded terms. For example, the 'concept of red' refers to the conceptual content expressed by 'red'.

---

[18] In order to bridge this gap one might think of theories as sums of sentences rather than sets. Even though this is metaphorical talk (if we take sums in the ordinary sense of the expression to be sums of individuals located in space and time), it captures the idea that the whole does not belong to a kind of entities possessing relevantly different properties from the properties possessed by the elements, parts or constituents.

[19] For a detailed discussion of how 'the property of being_', '_ness' and the like function, see Schnieder (2006).

2.5 Conclusion

Reduction reconciles diversity and directionality with strong unity. We now have a better idea of what strong unity amounts to, and we have an idea of how the guiding question should be addressed:

Q1: How can we reconcile *diversity* and *directionality* with strong *unity*?
Th. 1: *Strong unity* is the unity of identity. *Method*: To answer Q1 is to come up with an *explication* of the concept of reduction.

Unity is the unity postulated by a monist who is a realist about objects at "higher levels". Moreover, building on the model of explication proposed here, we are now in a position to start a discussion of the puzzle of reduction. A discussion of the puzzle suggests that diversity in reduction is conceptual in nature: Reduced and reducing item differ in how they are presented in a true reduction statement.

Bibliography

Balzer, Wolfgang. 1984. On the comparison of classical and special relativistic space-time. In *Reduction in science: Structure, examples, philosophical problems*, ed. Wolfgang Balzer, David A. Pearce, and Heinz-Jürgen Schmidt, 331–357. Dordrecht: Reidel.

Bechtel, William. 1994. Levels of descriptions and explanation in cognitive science. *Minds and Machines* 4: 1–25.

Bechtel, William, and Robert Richardson. 1993. Emergent phenomena and complex systems. In *Emergence or reduction? Essays on the prospects of nonreductive physicalism*, ed. Ansgar Beckermann, Hans Flohr, and Jaegwon Kim, 257–288. Berlin: de Gruyter.

Bickle, John. 2008. Real reduction in real neuroscience: Metascience, not philosophy of science (and certainly not metaphysics!). In *Being reduced*, ed. Jakob Hohwy and Jesper Kallestrup, 34–51. Oxford: Oxford University Press.

Block, Ned. 1997. Anti-reductionism slaps back. *Philosophical Perspectives* 11: 107–132.

Butterfield, Jeremy. 2011a. Emergence, reduction and supervenience: A varied landscape. *Foundations of Physics* 41: 920–959.

Butterfield, Jeremy. 2011b. Less is different: Emergence and reduction reconciled. *Foundations of Physics* 41: 1065–1135.

Carnap, Rudolf. 1934. *The unity of science*. London: Kegan Paul, Trench, Trubner, and Co.

Carnap, Rudolf. 1950. *Logical foundations of probability*. Chicago: University of Chicago Press.

Carnap, Rudolf. 1988 (1956). *Meaning and necessity. A study in semantics and modal logic*. Chicago: The University of Chicago Press.

Chalmers, David. 1996. *The conscious mind*. Oxford: Oxford University Press.

Churchland, Paul M. 1981. Eliminative materialism and the propositional attitudes. *The Journal of Philosophy* 78: 67–90.

Churchland, Paul M. 1985. Reduction, qualia and the direct introspection of brain states. *Journal of Philosophy* 82: 8–28.

Churchland, Patricia. 1986. *Neurophilosophy*. Cambridge, MA: MIT Press.

Craver, Carl. 2005. Beyond reduction: Mechanisms, multifield integration and the unity of neuroscience. *Studies in the History and Philosophy of Biological and Biomedical Sciences* 36: 373–395.

Crook, Seth, and Carl Gillet. 2001. Why physics alone cannot define the 'physical'. *Canadian Journal of Philosophy* 31: 333–360.

Davidson, Donald. 1969. The individuation of events. In *Essays in honor of Carl G. Hempel*, ed. Nicholas Rescher, 216–234. Dordrecht: Reidel.

Davidson, Donald. 1970. Events as particulars. *Noûs* 4: 25–32.

Dizadji-Bahmani, Foad, Roman Frigg, and Stephan Hartmann. 2010. Who is afraid of Nagelian reduction? *Erkenntnis* 73: 393–412.

Endicott, Robert. 1998. Collapse of the new wave. *The Journal of Philosophy* 95: 53–72.

Esfeld, Michael, and Christian Sachse. 2007. Theory reduction by means of functional sub-types. *International Studies in the Philosophy of Sciences* 21: 1–17.

Fodor, Jerry A. 1974. Special sciences: Or the disunity of science as a working hypothesis. *Synthese* 28: 97–115.

Fodor, Jerry A. 1981. *Representations: Philosophical essays on the foundations of cognitive science*. Cambridge, MA: MIT Press.

Hanna, Joseph. 1968. An explication of 'explication'. *Philosophy of Science* 35: 28–44.

Hellman, Geoffrey P., and Frank W. Thompson. 1975. Physicalism: Ontology, determination, and reduction. *Journal of Philosophy* 72: 551–564.

Horst, Steven. 2007. *Beyond reduction: Philosophy of mind and post-reductionist philosophy of science*. Oxford: Oxford University Press.

Kemeny, John G., and Paul Oppenheim. 1956. On reduction. *Philosophical Studies* 7: 6–19.

Kim, Jaegwon. 1976. Events as property exemplifications. In *Action theory*, ed. Douglas Walton and Myles Brand, 159–177. Dordrecht: Reidel.

Kim, Jaegwon. 1992. Multiple realization and the metaphysics of reduction. *Philosophy and Phenomenological Research* 52: 1–26.

Kim, Jaegwon. 1993. *Supervenience and mind. Selected philosophical essays*. Cambridge: Cambridge University Press.

Kim, Jaegwon. 2006. *Philosophy of mind*. Boulder: Westview Press.

Lewis, David. 1972. Psychophysical and theoretical identifications. *Australasian Journal of Philosophy* 3: 249–258.

Lewis, David. 1986. Events. In *Philosophical papers*, ed. David Lewis, 241–269. Oxford: Oxford University Press.

Melnyk, Andrew. 2003. *A physicalist manifesto: Thoroughly modern materialism*. Cambridge: Cambridge University Press.

Menzies, Peter. 1988. Against causal reductionism. *Mind* 98: 551–574.

Moulines, C. Ulises. 1984. Ontological reduction in the natural sciences. In *Reduction in science: Structure, examples, philosophical problems*, ed. Wolfgang Balzer, David A. Pearce, and Heinz-Jürgen Schmidt, 51–70. Dordrecht: Reidel.

Nagel, Ernest. 1961. *The structure of science. Problems in the logic of explanation*. New York: Harcourt, Brace & World, Inc.

Neurath, Otto. 1959. Sociology and physicalism. In *Logical positivism*, ed. Alfred J. Ayer, 282–317. New York: Free Press.

Nickles, Thomas. 1973. Two concepts of intertheoretic reduction. *The Journal of Philosophy* 70: 181–201.

Papineau, David. 1993. *Philosophical naturalism*. Oxford: Blackwell.

Papineau, David. 2002. *Thinking about consciousness*. Oxford: Oxford University Press.

Putnam, Hilary. 1975. On properties. In *Mathematics, matter and method. Philosophical papers*, ed. Hilary Putnam, 305–322. Cambridge: Cambridge University Press.

Quine, Willard V.O. 1960. *Word and object*. Cambridge, MA: MIT Press.

Quine, Willard V.O. 1986. *Philosophical logic*. Cambridge, MA: Harvard University Press.

Richardson, Robert C. 1979. Functionalism and reductionism. *Philosophy of Science* 45: 533–558.

Richardson, Robert C. 1982. How not to reduce a functional psychology. *Philosophy of Science* 48: 125–137.

Sarkar, Sahotra. 1992. Models of reduction and categories of reductionism. *Synthese* 91: 167–194.

Schnieder, Benjamin. 2006. Canonical property designators. *American Philosophical Quarterly* 43: 119–32.

Sklar, Lawrence. 2003. Dappled theories in a uniform world. *Philosophy of Science* 70: 424–441.

Smart, John J.C. 1959. Sensations and brain processes. *Philosophical Review* 68: 141–156.
Sneed, Joseph. 1971. *The logical structure of mathematical physics*. Dordrecht: Reidel.
Spurrett, David, and David Papineau. 1999. A note on the completeness of 'physics'. *Analysis* 59: 25–28.
Strawson, Peter F. 1963. Carnap's views on constructed systems versus natural languages in analytic philosophy. In *The philosophy of Rudolf Carnap*, ed. A. Schilpp, 503–518. La Salle: Open Court.
van Fraassen, Bas. 1980. *The scientific image*. Oxford: Clarendon.
Van Gulick, Robert N. 1992. Nonreductive materialism and the nature of intertheoretical constraint. In *Emergence or reduction? Essays on the prospects of non-reductive physicalism*, ed. Ansgar Beckermann, Hans Flohr, and Jaegwon Kim, 157–179. Berlin: de Gruyter.
Van Gulick, Robert N. 2001. Reduction, emergence and other recent options on the mind/body problem: A philosophic overview. *Journal of Consciousness Studies* 8: 1–34.
van Riel, Raphael. 2011. Nagelian reduction beyond the Nagel-model. *Philosophy of Science* 78: 353–375.
Waters, Kenneth. 1990. Why the anti-reductionist consensus won't survive. The case of classical Mendelian genetics. *PSA: Proceedings of the Biennial Meeting of the Philosophy of Science Association* 1: 125–139.
Wimsatt, William C. 1976. Reductive explanation: A functional account. In *PSA: Proceedings of the biennial meeting of the Philosophy of Science Association,* 1974, ed. G. Pearce, A.C. Michalos, C.A. Hooker, and R.S. Cohen, 671–710. Dordrecht: Reidel.
Wimsatt, William C. 2006. Aggregate, composed, and evolved systems: Reductionistic heuristics as means to more holistic theories. *Biology and Philosophy* 21: 667–702.
Witmer, Gene. 2008. (Review) Beyond reduction: Philosophy of mind and post-reductionist philosophy of science. *Notre Dame Philosophical Reviews*. http://ndpr.nd.edu/news/23449-beyond-reduction-philosophy-of-mind-and-post-reductionist-philosophy-of-science/. Accessed 5 July 2013.

Chapter 3
The Puzzle of Reduction

This chapter addresses the puzzle mentioned in the introduction, paving the way for the explication of the notion of reduction proposed in Chaps. 4 and 5.[1] Recall: The concept of reduction, as understood here, reconciles diversity and directionality with strong unity, without relying on elimination. But how can we reconcile strong unity, i.e. unity that is based on identity, with directionality? Identity is symmetric and reflexive. Directionality seemingly goes together with asymmetry and irreflexivity. Hence, the concept of reduction, so characterized, appears to be inconsistent. I propose a solution to this problem according to which '_reduces to_' generates hyper-intensional contexts and, thus, is sensitive to the conceptual or descriptive contents under which the reduced or reducing item is presented.[2] We thus arrive at

---

[1]This chapter builds upon ideas first presented in (van Riel 2008). These ideas were further developed in van Riel (2010, 2012). A somewhat similar thesis has recently been proposed by Jenkins (2011).

[2]Sensitivity to conceptual presentation has not loomed large in the philosophy of science-debate on reduction. There are some exceptions, or so it seems. Fritz Rohrlich incorporated a conceptual aspect in his approach to reduction (Rohrlich 1988). The basic idea of his is that theories at different levels differ in that they have different *cognitive values*. According to Rohrlich, at a higher level, we can speak of 'cognitive emergence' of objects in the sense that these objects "[...differ] perceptively in essential ways from anything that there was at any of the earlier stages [...]" (Rohrlich 1988, 298). This cognitive emergence is mimicked by the need for different conceptual structures. Unfortunately, Rohrlich is silent about how to conceive of these conceptual differences. The following suggestion makes it hard to uncover how conceptual structure is supposed to differ from ontological structure: "The two ontologies are associated with different cognitive levels [which, in turn, depend on different conceptual schemes, RvR]. On each level the theory gives us a different 'model of the world' which has *characteristic features that cannot be expressed by any other level* [...]" (Rohrlich 1988, 310). He explicitly argues in favor of *plural* ontologies. More straightforwardly, Ager et al. (1974) argue that we should conceive of the relation between temperature and mean kinetic energy in terms of identity, though the expressions used ('temperature' and 'mean kinetic energy') as well as the concepts expressed (the concept of temperature and the concept of mean kinetic energy) are different (Ager et al. 1974, 125 f.). They explicitly describe the field of the kinetic theory of gases as the field that tries to describe its target systems in terms of a certain list of properties (Ager et al. 1974, 123). This mimics, to a certain

R. van Riel, *The Concept of Reduction*, Philosophical Studies Series 121,
DOI 10.1007/978-3-319-04162-9_3,

diversity: It is conceptual or descriptive diversity, which goes together with unity in that one and the same entity, say, pain, is presented under different concepts, say, the concept of pain and the concept of C-fiber firing. The discussion of a rival solution, according to which diversity stems from differences in reduced and reducing theories will have to wait until Chap. 8.

3.1 The Puzzle

The puzzle can be given different forms. Up to now, we did rely on an intuitive understanding of it, which is connected to an intuitive understanding of "directionality". Here is another schematic version of the puzzle, a version that appeals to *Leibniz Law*: Assume that a reduces to b. Therefore, $a = b$. However, by *Leibniz Law*, for any property x, if a exemplifies x then b exemplifies x (since $a = b$). a exemplifies the property of reducing to b. Hence, b exemplifies the property of reducing to b. Nothing reduces to itself. Hence, the concept leads into contradiction. This is a disturbing result.

It is even more disturbing since it can be cast into terms that perfectly match the prose in the reduction debate: How is identity-based reduction supposed to be compatible with the characterization of reduction in terms of *levels*, organized *hierarchically*? Schematically: If a reduces to b, then $a = b$. Assume that if a reduces to b, then a is at a higher level than b (this is the very idea of reductive hierarchies). How can this be, if $a = b$, and entities are located at one reductive level within a hierarchy only? Again, the concept leads into contradiction. So, why shouldn't we, in an appropriate explication of the concept of reduction, get rid of directionality in the first place? We should not, I submit, because directionality is *important*. It is reasonable to assume that for any attempt to explicate some concept C, if C is not confused or incoherent, and if C is used in some explanatory context E, the explication of C should be equally explanatory in the context of E. It should be clear that a good explication of reduction should yield a definition of a directional relation

degree, the idea pursued here. However in (Ager et al. 1974) this idea is left sketchy. The authors have a different target, namely, the alleged necessity of bridge-laws for *reductions* (that is: the role of bridge-laws is discussed in the context of achieving reductions, rather than in the context of some reduction relation which holds independent of whether or not the reduction is actually carried out). Note that this idea is not to be conflated with the idea that different representational schemes amount to different explanatory links and hierarchies – as long as these are spelled out in purely ontological terms. Sarkar (1998), Bechtel and Richardson (1993), and Wimsatt (1974) describe the idea of different representational schemes, which are individuated via the relations they refer to within the system to be represented (for example, functional vs. structural relations in mechanistic explanations). There is also one strand in the philosophy of mind that can be illuminated building on this notion of reduction: The qualia-problem has been stated in a number of different forms (Jackson 1982, 1986; Davidson 1986; Shoemaker 1982). One prominent reaction to this problem roughly runs as follows: Difference in conceptualization does not imply a difference in ontology (Levine 1998; Papineau 2002, 2007; Tye 1999). Thus, one might suggest that qualia-reduction requires a difference in conceptual presentation and, at the same time, identity of properties.

as well. Directionality is central to the role the notion plays in explanations of a hierarchy of sciences, in the characterization of positions of ontological monism, and of particular reductive statements; claiming, say, that water reduces to H_2O, or that temperature reduces to mean kinetic energy, we are thus committed to the claim that H_2O does not reduce to water, and that mean kinetic energy does not reduce to temperature. If we cannot account for the directionality, we lose a great deal of the interesting features of the concept of reduction. That high-level sciences possibly reduce to fundamental science is a substantial point only if in virtue of the fact that the relevant link is reductive, it is not the case that the fundamental science reduces to the high-level sciences. Accordingly, the concept of reduction pertinent in the pre-theoretical descriptions is used to pick out a directional relation between items at different levels. Thus, identity cannot be sufficient for reduction. So, does the intuitive characterization of reduction, that reduction is supposed to reconcile directionality with strong unity, ultimately rest on a basic confusion? Fortunately, it does not. There is an almost obvious solution to the puzzle.

3.2 Hyper-intensionality

In (van Riel 2008, 2010, 2012) I have argued that a statement of the form '*a* reduces to *b*' expresses a truth, if it does, partly in virtue of semantic facts about the relevant instances of '*a*' and '*b*' *other* than facts about reference, that '_reduces to_' generates hyper-intensional contexts, and that, therefore, the reduction relation is not a binary relation. In what follows, I will briefly summarize the main argument for the claim that the reduction-predicate generates hyper-intensional contexts. Hyper-intensionality for operators that take designating expressions, such as terms or descriptions, as arguments is here understood as follows:

(Def. Hyper-Intensionality): x generates hyper-intensional contexts iff

- in at least one of the positions in which it may take an argument, substitution *salva veritate* of co-intensional expressions is not always possible, and
- substitution *salva veritate* of synonymous expressions is always possible.

Note that in the present context, the relevant set of co-intensional expressions will be co-referential or co-designating terms *a*, *b*, such that if '$a = b$' expresses a truth, then so does 'necessarily, a = b'.[3] For what follows, I will thus switch between 'co-intensional', 'co-designating', and 'co-referential'.

[3] If there are contingent identities that are metaphysically contingent under every possible reading, and if such identity statements play a role in reduction, the account would have to be slightly adjusted, so as to capture a merely intensional interpretation of '_reduces to_'. This would, I submit, best be captured by distinguishing between a weak and a strong version of identity-based reduction – the former merely requiring contingent identities, whereas the latter requires necessary

In the literature on hyper-intensionality, it is not referred to substitutability *salva veritate* of synonymous expressions. However, to give an interpretation of the *relata* of the reduction relation later, mentioning of this aspect is important, as will become apparent below. Here is the argument:

(P1) If a predicate of the form '_*R*_' does not allow for substitution *salva veritate* of co-intensional expressions in its argument-positions, but if it does allow for substitution *salva veritate* of synonymous expressions, then the predicate generates hyper-intensional contexts (on its left hand side as well as on its right hand side).

(P2) Sentences of the form '*a* reduces to *b*' allow for substitution *salva veritate* of synonymous expressions in the positions of '*a*' and '*b*' (in general).

(P3) Sentences of the form '*a* reduces to *b*' do not (in general) allow for substitution *salva veritate* of co-intensional expressions in the positions of '*a*' and '*b*'.

(C) Therefore, '_reduces to_' generates hyper-intensional contexts (on its left hand side as well as on its right hand side).

(P1) follows from the definition of hyper-intensionality just given. As for (P2): Assume that 'snake' and 'serpent' are synonymous (in the sense that x is synonymous to y iff either x and y are co-referential and lack meaning, or x and y have the same meaning), and that 'the kind snake reduces to the kind [some appropriate expressions in terms of DNA]' expresses a truth. Then, 'the kind serpent reduces to the kind [some appropriate expressions in terms of DNA]' will express a truth as well. This seems to be independent of whether or not 'snake' and 'serpent' directly refer, or express a conceptual content. Moreover, there is nothing special in 'snake' and 'serpent'; thus, we can generalize to (P2). As for (P3): If water reduces to H_2O, then H_2O does not reduce to water! Reduction is directional. By assumption, water = H_2O. Hence, (P3). Thus, in a sense, the puzzle wears its solution in its sleeve.[4]

But isn't this puzzling? Consider the following line of thought: If reduction is tied to identity so that if a reduces to b, then $a = b$, then reduction reflexive, transitive and symmetric (see, for example, Trogdon 2013, Section 6). It seems that the best reason for assuming that reduction is symmetric and reflexive roughly runs as follows: the fact that (if a reduces to b then $a = b$) *implies* that reduction has the structural features of identity.

---

identities. The former could accommodate the assumption that a statue reduces to the clay it is formed of, but that this is merely a contingent identity.

[4]Note that this has an interesting consequence for the application of Leibniz' Law: Just because 'a exemplifies F-ness' expresses a truth, and 'b exemplifies F-ness' expresses a falsehood, it does not follow that a $\neq$ b.Schnieder (2006c) has made a similar point about pragmatic aspects of substitution of co-referential expressions. The point made here is semantic in nature, as will become apparent below. Independent of considerations about Leibniz law, a pragmatic interpretation of substitution failures would blur the distinction. Such a strategy has been pursued by Recanati in his (2000), concerning substitution failures in belief-contexts.

Brief reflection on how the arity of a predicate that generates hyper-intensional contexts is connected to the arity of the relation or property it expresses will help solving this problem: It can easily be shown that if the reduction predicate generates hyper-intensional contexts then this inference is not licensed. Since the hyper-intensional reading perfectly fits the way the predicate is used in philosophical discourse, we should resist what I suggested is a mistaken inference – the inference from the observation that if *a* reduces to *b* then $a = b$, to the thesis that reduction has the formal features of identity. Here is an argument for the claim that if the arity of a predicate that generates hyper-intensional contexts is of arity n, then the relation it expresses is of an arity $m \neq n$. Correspondingly, the reduction relation, being expressed by a binary hyper-intensional predicate, is not a binary relation. Therefore, it is neither symmetric nor reflexive (nor asymmetric, nor irreflexive). Moreover, it will be suggested that relations expressed or expressible by predicates that generate hyper-intensional contexts in the sense specified above are sensitive to semantic facts about the expressions occurring within their scope *other* than facts about reference or designation. The next section is a little bit technical in nature. So, it can be skipped by those who are not interested in the details, and who are willing to take for granted that '_reduces to_' is, in this sense, sensitive to semantic facts about its arguments other than those regarding reference or designation.[5]

3.3 Hyper-intensional Predicates: Counting Arguments vs. Counting Relata

Before giving a detailed version of the argument, let me put it roughly as follows: Hyper-intensionality amounts to the fact that we cannot always substitute co-referential expressions *salva veritate*. If the arity of the relation expressed by a predicate is the same as that of the predicate, *n*, then the following conditional is true: If a sentence, which is obtained from a predicate *P* by adequately substituting *n* singular terms in the argument position, is true then these terms single out an n-tuple of referents of these terms that is mapped onto the value TRUE (and no other truth-value) by the relation expressed by *P*,[6] and onto FALSE (and no other truth-value) if the sentence is false. For hyper-intensional expressions, this is not the case. Therefore, the arity of a relation expressed by a hyper-intensional expression is not identical to the arity of that expression. (For what follows, I will

[5]An independent argument has been given in van Riel (2010). Jenkins (2011) suggests this interpretation, arguing that alternatives (namely, pragmatic ones) do not seem as appropriate as this one.

[6]Alternatively, you could replace this premise and talk about the relation between a sentence's being true and the corresponding tuple of objects belonging to the relevant extension (across worlds). Or, even simpler, that precisely this tuple has the property if the sentence is true, and does not have the property if the sentence is false, and nothing has and does not have the property at the same time.

talk about 'reference' only, assuming that an analogous argument could be run for other semantic relations that are such that they appropriately relate an object to an expression, thereby enabling us to predicate something of it.)

Note that to keep things simple, I will ignore modal aspects. In particular, it will be convenient to conceive of properties as functions from individuals or tuples of individuals to truth-values. In a similar spirit, I will restrict quantification to actually existing predicates and sentences (predicates and sentences that have been or can be generated using an actual natural language).[7] Here is the argument in a more structured format. In order to make the argument short, we need a notion of a specific procedure, which corresponds to successful tests for hyper-intensionality – that of substituting a singular term in a true sentence for a co-intensional term, thereby rendering the sentence false. It will be convenient to conceive of this as follows:

(Definition 'Hyper Substitution'): In a sentence *s*, containing a singular term *y*, we *hyper-substitute* $y*$ for *y* iff

- *s* is true
- *y* is co-intensional with $y*$, and
- this substitution turns the sentence into a sentence that expresses a falsehood.

As a convention, sentences that can be obtained by hyper-substitution from a sentence *s* will be referred to as '$variants_{HS}$ of *s*'. An interesting feature of hyper-substitution is this:

(Principle Hyper-Substitution): For any predicate *r*, if *r* generates hyper-intensional contexts, then there is at least one sentence *s*, one sentence s', such that *s* is true, and *s* is obtained from *r*, and s' is obtained by hyper-substitution from *s*.[8]

This principle just states that whenever we have hyper-intensionality, we have possible hyper substitution. Why is this so? Because hyper-intensionality implies the possibility of hyper substitution – an expression generates hyper-intensional contexts iff in one of its argument-positions, we can, at least in principle, substitute an expression for a co-referential expression and thereby turn a true sentence into a

---

[7] I am confident that a modal version of the argument could be given. Moreover, I will ignore sentences in which we quantify into an argument position, which is within the scope of a hyper-intensional predicate.

[8] This principle hinges on the assumption that no hyper-intensional predicates are such that they have true or false instances only; if there are, we need to go modal and hope that there are no hyper-intensional predicates that necessarily have true or false instances only. Cases like '$2+2=5$ and _reduces to_' can probably be coped with, by requiring that any argument be erased, or be the possible candidate for substitution, or by requiring predicates to occur in atomic sentences. Moreover, it is assumed that for any hyper-intensional predicate, there is, at least in principle, a test for hyper-intensionality.

false one; that is: if we can hyper-substitute. I will rely on an intuitive understanding of what it is to obtain a sentence from a predicate – we do so by closing the open sentence adding the appropriate number of arguments.

Now, we need an idea of how the singular terms of a sentence of the form '$Ra_1 \ldots a_n$' relate to the tuples taken as arguments by the function corresponding to (or being) the relation expressed by '$Rx_1 \ldots x_n$'. Let us say that the string of singular terms (the singular terms, for short) of '$Ra_1 \ldots a_n$' *single(s) out* the tuple $<a_1, \ldots a_n>$. More precisely, the singular terms of a sentence s single out the tuple x iff the first term occurring in s refers to the first element of the tuple, the second term refers to the second element of the tuple, and so forth.[9] As a matter of convenience, we will thus assume that unary relations (or properties) take n-tuples as arguments, with $n = 1$.

From the notion of hyper-substitution and the idea of a sentence's singular terms singling out a tuple, the following principle can be obtained:

(Principle Singling Out Tuples): For any two sentences, s, s', if s' is a variant$_{HS}$ of s, then the singular terms of s and s' single out the same tuple.

The singular terms of two sentences single out the same tuple iff the terms occurring in the sentences are, in the appropriate order, co-intensional. The following two principles reflect the idea that properties and relations are (or uniquely determine) functions, which map individuals or, here, tuples of individuals onto truth-values. The relation between the arguments of a predicate and the function expressed by that predicate seems to be governed by the following three principles:

(Principle Mapping I): For any relation r^*, if r^* is the relation expressed by an n-ary predicate r, and r^* is of arity n, then, for any sentence s, for any n-tuple x, if s is true, and s is obtained from r, and the singular terms of s single out x, then r^* maps x onto TRUE.

(Principle Mapping II): For any relation r^*, if r^* is the relation expressed by an n-ary predicate r, and r^* is of arity n, then, for any sentence s, for any n-tuple x, if s is false, and s is obtained from r, and the singular terms of s single out x, then r^* maps x onto FALSE.

(Principle Mapping III): No relation expressed by a predicate maps a tuple onto the values TRUE and FALSE at the same time.

---

[9]What about sentences that contain terms seemingly referring to fictional entities? This is a more general problem, and whatever the correct solution might be, there seems to be no reason to believe that it cannot be accommodated within the present framework. Similarly, I will ignore possible hidden indexicals concerning contexts or the actual world; then, these would add to the arity of the relation. One prominent interpretation of hyper-intensionality is itself a variant of a hidden-indexical approach (see, for example, Forbes 2006). The argument presented here can be seen as showing that something along these lines must be correct for hyper-intensional expressions.

Principle Mapping III is, I guess, beyond a doubt (even if you drop the principle of bivalence): No function expressed by a predicate of natural language takes its arguments to the values TRUE and FALSE at the same time. Principle's Mapping I and II seem equally intuitive: The very idea of the conception of properties as functions is that a property maps the arguments delivered by the singular terms of the sentence onto the value TRUE iff the sentence expresses a truth, and to the value FALSE iff the sentence expresses a falsehood (again, this seems to be acceptable even for the intuitionist). Here, an even weaker (and, as it will turn out, more appropriate) assumption is made: This will hold (at least) if the arity of the relation expressed by a predicate is identical to the arity of the predicate.

Now, let R be a predicate with n argument positions that generates hyper-intensional contexts in at least one argument position, let R^* be the relation expressed by R, let S be a true sentence obtained from R, let S' be a variant$_{\mathrm{HS}}$ of S, and let Y be the n-tuple singled out by the singular terms of S and S' respectively. *(Principle Singling Out Tuples)* and *(Principle Hyper-Substitution)* together guarantee that whenever we have a hyper-intensional predicate, we will have a true sentence obtained from the predicate, a variant$_{\mathrm{HS}}$ thereof, and a tuple singled out in this specific way.

Now, to show that this leads into inconsistency, we assume that

(P1)	R^* is of arity n.	
(P2)	Y is mapped onto TRUE by R^*.	(From *Principle Mapping I*, the assumption that R^* is of arity n, the assumption that S is true and the singular terms of S single out Y.)
(P3)	Y is mapped onto FALSE by R^*.	(From *Principle Mapping II*, the assumption that R^* is of arity n, the assumption that S' is false and the singular terms of S' single out Y.)
(P4)	Therefore, Y is mapped onto TRUE and FALSE by R^*.	(From (P2)-(P3).)
(P5)	(P4) contradicts *Principle Mapping III.*	
(P6)	Therefore: R^* is not an n-ary relation.	

(P6) contradicts (P1). Nothing hinges upon 'R', R^*, S, S', or Y. Hence, we can generalize: If a predicate generates hyper-intensional contexts, then the arity of the relation it expresses is not identical to the arity of the predicate. Correspondingly, if the reduction-predicate generates hyper-intensional contexts, then the reduction relation is not a binary relation. Therefore, it is neither asymmetric, nor symmetric, nor irreflexive, nor reflexive. So, possibly, sentences of the following form may be true at the same time: 'a reduces to b', '$a = b$', and 'b does not reduce to a'. Moreover, the fact that an instance of 'a reduces to b' implies the corresponding instance of '$a = b$' does not force us to accept that reduction has the structural features of identity.[10] Let us thus turn to the relation between a hyper-intensional predicate and the relation it signifies.

We have shown that the arity of the relation expressed by a hyper-intensional predicate is not identical to the arity of the predicate. But what, then, *is* the arity of a relation expressed by an n-ary predicate? My suggestion is this: The arity of a relation of a hyper-intensional n-ary predicate that generates hyper-intensional contexts in m argument positions is $n + m$. And what are the *relata*? It is (i) the referents (or designata) of these expressions and (ii) the conceptual contents expressed by these expressions (or, if they lack conceptual content, a semantic fact about their lacking a conceptual content).

Recall that the truth-value of sentences of the form '*a* reduces to *b*' is insensitive to substitution of synonymous expressions – no matter what we substitute for '*a*' or '*b*', as long as it is synonymous with what we substitute it for, we will not alter the truth-value of the sentence. Assume that synonymy is sameness of meaning or sameness of conceptual content (such that necessarily, directly referring expressions which are co-referential express the same conceptual content). Then, the sentence's truth-value partly depends upon the meanings (or facts about the meanings) of the expressions that flank the reduction-predicate, not only on the expression's referents or designata. Assume that this generalizes (and this seems plausible – there is, at the face of it – nothing special in '_reduces to_'). If so, for any argument of a predicate that generates hyper-intensional contexts in this argument position, we get two entities relevant for a corresponding sentence's truth-value, the referent *and* the conceptual content. Thus, for an n-ary predicate with m argument-positions in which it generates hyper-intensional contexts, we get n referents and m conceptual contents (or facts about the expression's meaning) and, hence, $n + m$ objects that bear upon whether or not the sentence expresses a truth or a falsehood, that is $n + m$ objects the function expressed by the predicate takes to a truth-value.

3.3.1 An Objection

Now, there might be a Frege-inspired counter-argument.[11] Assume that, within the scope of hyper-intensional expressions, expressions shift their reference to their

---

[10] This supports Jenkin's (2011) point: From the fact that an instance of '*a* is grounded in *b*' implies the corresponding instance of '*b* is not grounded in *a*', it does not follow, without further ado, that the grounding relation is asymmetric.

[11] I would like to thank Marta Campdelacreu for pointing out this argument to me, during a discussion of this argument in a Logos-Seminar in Barcelona in 2012. Jenkins (2011) discusses a somewhat similar alternative (without adopting it): That the semantic values of the terms occurring in the scope of hyper-intensional expressions may designate tuples of objects, one element of which is the ordinary object signified or referred to by the expression. This view will yield an equivalent result; we have to decide between postulating systematic reference shift and postulating that systematically, the arity of some predicates does not reflect the arity of the relation they express.

conceptual contents, so that the Principle about singular terms singling out tuples given above is mistaken. Here are three reasons for why I think that this is an idea we are better off without in the present context. First: Assume that there are directly referential expressions. Within hyper-intensional contexts, they would lack a referent. At least some kind-terms are good candidates for being directly referential. They are, at the same time, promising candidates for occurring on the left-hand side of the reduction-predicate in true reduction statements. These sentences would turn out false, or fall into a truth-value gap. Even if we assume that hyper-intensional expressions change their function in the case of directly referential expressions, such that these expressions keep their ordinary function, we would have the odd result that sentences such as 'water reduces to H_2O' state that a kind (water) reduces to the concept of H_2O. This brings us to the second point:

It might be reasonable to assume that a sentence such as 'Peter believes that snow is white' primarily says something about (i) Peter and (ii) the proposition that snow is white (i.e. the conceptual content expressed by 'snow is white'); within the Fregean picture, reference-shift is responsible for this result, and the relevant *relata* of the belief-relation are persons (what is referred to or designated on the left hand side of 'believes (that)') and propositions (what is expressed and not referred to in ordinary contexts on the right hand side of 'believes (that)'). For other cases, such as *intends to*, *reduces to* etc. the consequences seem, again, rather odd. Intending to drink a beer, I do not intend a conceptual content. And if water reduces to H_2O, then we do not primarily say something about the concept of water and the concept of H_2O. An appropriate explication should reflect these intuitions (or explain them away). Relying on the strategy of reference-shift, we would be unable to capture these intuitions, and it is not immediately clear how we could explain them away.

I think that a good idea to paraphrase reduction statements is to say that, in the case of water and H_2O, water, when presented *as* water, reduces to water/H_2O, when presented *as* H_2O.[12] Turning back to the intuitive characterization of reduction: This captures the idea that reduction reconciles diversity and directionality with strong unity. If we could come up with an interpretation along these lines, we should adopt it. The task is *explication* of a technical concept. Even if 'reduction' generates hyper-intensional contexts of an uncommon kind, in the present context, there is no reason to fear this result. Thus, I think that we are safe to conclude that the argument developed above is not a non-starter.[13] So, we are now in a position to come up with a solution to the puzzle and point to where we should look for a full-blown explication of the concept of reduction.

---

[12]For an interpretation of phenomena of hyper-intensionality along these lines, see (Forbes 2006). Note that, depending on how one interprets Recanati's reference to 'pragmatic' aspects, his suggestion for treating the functioning of 'beliefs' is similar (Recanati 2000).

[13]Here is one general remark: Given these problems for an application of the Frege-picture to the case of reduction, one might feel tempted to ask in response: Why should 'believes (that)' function differently? What would be the price to pay when 'Peter beliefs that snow is white' should be paraphrased as 'Peter believes (the fact or state of affairs) that snow is white *as* presented in that specific way'?

3.4 A Solution to the Puzzle

Reduction reconciles diversity with strong unity as follows: the arguments of the reduction predicate in true reduction statements differ with respect to some semantic properties they have (thus, we get diversity). They do not differ with respect to reference or designation or intension (thus, we get strong unity). Thus, the reduction-relation is, despite appearances, not asymmetric. Asymmetry is defined for binary relations. Reduction is not a binary relation. Therefore, it is not irreflexive either. Nor is it reflexive, nor transitive. Building on the idea that the sentence 'water, when presented as water reduces to water, when presented as H_2O' gives a good idea of the logical structure of reduction statements of the form 'water reduces to H_2O', we can define a number of cognate notions. This can be done in the spirit of Schnieder (2010) who defines a notion of asymmetry for 'because' that does not build upon the idea that 'because' expresses a relation. Here is the motivation and the basic idea:

> However, that 'because' expresses a relation is a potentially controversial claim about its semantic function (more on it below). Since not every philosopher who classifies 'because' as asymmetrical and/or irreflexive would want to underwrite that view on the function of 'because', it is worthwhile looking for an alternative explication of classifying an operator as asymmetrical. Such an alternative can be framed in of certain inferential properties of the sentences involving the operator: an operator may, for instance, always allow the sentences it connects to be permutated salva veritate ('or' is a case in question, while 'if' is not). Because of this inferential property, a connective could be called symmetrical. A precise definition along these lines can be given by the aid of substitutional quantification. Let us define the substitution class for the variables 'p' and 'q' as comprising all sentences of the respective object-language L. (In order to avoid unwelcome results due to the limited expressive power of L, the variables should not only range over the actual expressions of L, but also over expressions in possible extensions of L.) Then we can define asymmetry of dyadic sentential connectives as follows:
>
> '$\rhd$' is <u>asymmetrical</u> $\leftrightarrow$ df. $\forall p\ \forall q$: $(p \rhd q) \rightarrow \neg (q \rhd p)$
>
> (The sign '$\rhd$' can be regarded as a substitutional variable whose substitution class are the two-place sentential connectives.)
> In the same vein, we can define irreflexivity for dyadic sentential connectives:
>
> '$\rhd$' is <u>irreflexive</u> $\leftrightarrow$ df. $\forall p$: $p \rightarrow \neg (p \rhd p)$ (Schnieder 2010, 321 f.)

Building on this idea, we could describe 'reduction' as <u>asymmetric</u> as well. We just change the substitution classes for our variables, from sentences to terms, and the variables, to avoid confusion. We thereby obtain a notion of asymmetry that is similar to the notion of directionality introduced above:

'**R**' is <u>asymmetrical</u> $\leftrightarrow$ df. $\forall a\ \forall b$: $(a\ \mathbf{R}\ b) \rightarrow \neg (b\ \mathbf{R}\ a)$[14]

---

[14]Note that this is, maybe, a misguided interpretation of the relevant sort of asymmetry. Consider a (very odd) language where meaning is at least partly determined by position in a sentence, so that in 'p because q', p has the meaning of q and q has the meaning of p in 'q because p'. Then, 'because' is not asymmetrical in the sense specified above. We should thus not work

Thus, interestingly, this sort of asymmetry is exhibited not only by expressions that (i) express binary relations and (ii) do not express relations at all, as, maybe, 'because', but also by (iii) expressions that express relations of an arity >2. Can we, in a similar spirit, obtain a definition for the relevant kind of irreflexivity? We cannot. The relevant sort of irreflexivity is not captured by the corresponding definition:

'**R**' is irreflexive $\leftrightarrow$ df. $\forall a$: $\neg(a\ \mathbf{R} a)$

To be sure, 'reduces to' is irreflexive in this sense; however, 'reduces to' is irreflexive in a more demanding sense as well. It does not allow for two *synonymous* expressions to occur in its argument-positions at the same time. We could thus define a notion of irreflexivity as follows:

'**R**' is irreflexive* $\leftrightarrow$ df. $\forall a, b$: *a is synonymous to b* $\rightarrow \neg(a\ \mathbf{R}\ b)$

In this sense, 'reduces to' is irreflexive. But isn't there a more 'ontic' interpretation of asymmetry and irreflexivity available for reduction? Here is a suggestion to define asymmetry for a quaternary relation, where the quantification is not substitutional:

R is asymmetrical$_{\text{ontic}}$ $\leftrightarrow$ df. $\forall x, x^*, y, y^*: x, y\ R\ x^*, y^* \rightarrow \neg\ x^*, y^*\ R\ x, y$.

The reduction-relation is asymmetric in this sense. It is a relation better expressed by the predicate '_ when presented under _ reduces to _ when presented under _'. Similarly, we could define an ontic version of irreflexivity for quaternary relations:

R is irreflexive$_{\text{ontic}}$ $\leftrightarrow$ df. $\forall x, y: \neg\ x, y\ R\ x, y$.

On each of these interpretations, neither appeal to Leibniz' Law, nor the assumption that if reduction-statements imply corresponding identity-statements, then reduction is symmetric, reflexive etc., shows that the concept of reduction, as conceived of here, is defective. We get diversity and strong unity, and thereby solved the puzzle.

3.5 Conclusion

The previous Chap. 2 offered an idea of what strong unity consists in, and it suggested that a definition of the concept of reduction should rely on an explication. This chapter suggested that diversity if conceptual in nature:

---

with substitutional quantification without, additionally, invoking conditions on the meanings of the relevant expressions. What does this show? It does not show that the characterization fails for English. But it shows that the characterization maybe does not capture the feature it is supposed to capture. It adequately models the "asymmetry" of 'because' and other expressions just because our English (and other natural languages) does not contain rules that allow for disambiguation in the way just sketched.

Q1: How can we reconcile *diversity* and *directionality* with strong *unity*?

...

Th. 2: *Diversity* is conceptual in nature, as a discussion of the puzzle of reduction shows. It is to be accounted for in terms of the conceptual contents of the expressions that pick out the reduced/the reducing entity in a true reduction statement.

...

'_reduces to_' generates hyper-intensional contexts. Diversity is to be cashed out in terms of a difference in semantic features of the expressions that flank the reduction predicate in true reduction statements. We are thus able to reconcile diversity with strong unity. But what about *directionality*?

Bibliography

Ager, Tryg A., Jerrold L. Aronson, and Robert Weingard. 1974. Are bridge laws really necessary? *Noûs* 8: 119–134.

Bechtel, William, and Robert Richardson. 1993. Emergent phenomena and complex systems. In *Emergence or reduction? Essays on the prospects of nonreductive physicalism*, ed. Ansgar Beckermann, Hans Flohr, and Jaegwon Kim, 257–288. Berlin: de Gruyter.

Davidson, Donald. 1986. Knowing one's own mind. *Proceedings and Addresses of the American Philosophical Association* 60: 441–458.

Forbes, Greame. 2006. *Attitude problems*. Oxford: Oxford University Press.

Jackson, Frank. 1982. Epiphenomenal qualia. *Philosophical Quarterly* 32: 127–136.

Jackson, Frank. 1986. What Mary didn't know. *Journal of Philosophy* 83: 291–295.

Jenkins, Carrie S. 2011. Is metaphysical dependence irreflexive? *The Monist* 94: 267–276.

Levine, Joseph. 1998. Conceivability and the metaphysics of mind. *Noûs* 32: 449–480.

Papineau, David. 2002. *Thinking about consciousness*. Oxford: Oxford University Press.

Papineau, David. 2007. Phenomenal and perceptual concepts. In *Phenomenal concepts and phenomenal knowledge*, ed. Torin Alter and Sven Walter, 111–144. Oxford: Oxford University Press.

Recanati, François. 2000. *Oratio Obliqua, Oratio Recta, an essay on metarepresentations*. Cambridge, MA: MIT Press.

Rohrlich, Fritz. 1988. Pluralistic ontology and theory reduction in the physical science. *The British Journal for the Philosophy of Science* 39: 295–312.

Sarkar, Sahotra. 1998. *Genetics and reductionism*. Cambridge: Cambridge University Press.

Schnieder, Benjamin. 2006. 'By, Leibniz's Law' – remarks on a fallacy. *Philosophical Quarterly* 56: 39–54.

Schnieder, Benjamin. 2010. A puzzle about 'because'. *Logique et Analyse* 53: 317–343.

Shoemaker, Sidney. 1982. The inverted spectrum. *Journal of Philosophy* 79: 357–381.

Trogdon, Kelly. 2013. An introduction to grounding. In *Varieties of dependence*, ed. Miguel Hoeltje, Benjamin Schnieder, and Alexander Steinberg, 97–122. München: Philosophia Verlag.

Tye, Michael. 1999. Phenomenal consciousness: The explanatory gap as cognitive illusion. *Mind* 108: 705–725.

van Riel, Raphael. 2008. From water to H_2O – What reduction is about. PhilSci archive: http://philsci-archive.pitt.edu/4101/1/Reduction%2Dpaper%2Donline.pdf. Accessed 5 Jan 2013.

van Riel, Raphael. 2010. Identity-based reduction and reductive explanation. *Philosophia Naturalis* 47–48: 183–219.

van Riel, Raphael. 2012. Identity, asymmetry, and the relevance of meanings for models of reduction. *The British Journal for the Philosophy of Science*. doi:10.1093/bjps/axs028.

Wimsatt, William C. 1974. Complexity and Organization. In *PSA: Proceedings of the biennial meeting of the Philosophy of Science Association*, 1972, *Boston studies in the philosophy of science*, ed. Kenneth Schaffner and R. Cohen, 67–86. Dordrecht: Reidel.

Chapter 4
Reductive Explanation and Mechanistic Explanation

Reduction reconciles diversity and directionality with strong unity, without relying on elimination. *Unity* is achieved by identification; *diversity* is conceptual or descriptive in nature; but what about *directionality*? Directionality can be accounted for in terms of explanation. The present chapter and the following chapter motivate this idea and connect it to the thesis that reductive diversity is conceptual diversity.

4.1 Reduction – A First Explication

Consider, again, our standard example: Water reduces to H_2O. I propose the following reconstruction of this sentence's truth-conditions: Water reduces to H_2O iff for every *x*, if *x* is water then (*x* is water *because* *x* is H_2O), and being water = being H_2O. We should add that no conceptual explanation is involved; otherwise, all conceptual explanations, what was briefly gestured at in Sect. 2.2.3, such as 'this is a drake because this is a male duck', would turn out to be reductive explanations. Mere identity of properties does not suffice for conceptual constitution or analysis, or whatever conceptual dependence consists in – we cannot account for the concept of water in terms of the concept of H_2O. Note that one may want to introduce a broader notion of reduction that encompasses both sorts of reduction, conceptual as well as non-conceptual reduction. These kinds of explanation may even bear striking similarities. Here, we are concerned with the non-conceptual case of reductive explanation only.[1] This yields a first explication of the notion of reduction, here restricted to the reduction of *types* or *kinds*.

[1]Intuitively, the form of reduction we are concerned with here is tied to *metaphysical* as opposed to *conceptual* dependence.

R. van Riel, *The Concept of Reduction*, Philosophical Studies Series 121,
DOI 10.1007/978-3-319-04162-9_4, © Springer International Publishing Switzerland 2014

(Explication I – Kind Reduction) *F-ness* reduces to *G-ness* iff

(i) for every *x*, if *x* is *F* then (*x* is *F because x* is *G*), and
(ii) *F-ness* = *G-ness*.

Condition (i) is, thus, to be interpreted as a *non-conceptual* explanation. Instances of it do not illuminate the concept expressed by the predicate in the position of '*F*'. Note again that the logical form of 'water reduces to H_2O' is maybe not perfectly captured by '*F-ness* reduces to *G-ness*' – one might expect a quantified phrase, rather than a phrase containing terms referring to properties or kinds. However, sentences such as 'water reduces to H_2O' are used to express the idea that the kind water is identical to and reduces to the kind H_2O; thus, deviating from what might seem to be the grammatical surface structure seems justified (to repeat: we are dealing with a technical term, where occasionally, the *intended* use weighs more than grammatical surface-structure).

This candidate explication suffices, in a sense, for a full-blown explication of the concept of (property-)reduction. It guarantees strong unity because of condition (ii). Note that due to this condition, we do not have to add a necessity operator in condition (i). Condition (i) captures directionality, and it ensures conceptual diversity: In condition (i), we face the interesting case where explanans and explanandum of an explanation are *intensionally equivalent*, but do not yield conceptual analysis. This can be illustrated referring to the idea of Russellian propositions: The Russellian proposition expressed by 'this is water' and 'this is H_2O', when uttered in an appropriate context that fixes the meaning of 'this', is, roughly, *<the object referred to by 'this', being water>* in both cases. The difference between explanans and explanandum in virtue of which we get explanatory dependence is *conceptual* in nature.[2]

This definition clearly distinguishes reduction from other dependence relations such as causation on the one hand, and conceptual dependence on the other. Causation is incompatible with condition (ii) – without an appropriate time-index, the properties alluded to in a causal explanation have to be distinct. One may want to hold that one is a smoker now *because* one was a smoker last year, and that this explanation can be given a causal interpretation. Removing reference to times, however, radically changes the situation. No one's being a smoker at t_1 *causally depends* on her being a smoker at t_1, in any sense. The relation of this and similar definitions to characterizations of ontological dependence or grounding will be

[2] Later, it will be argued that the idea that reduction is an explanatory relation or goes together with explanatory relations is wide spread: The early Nagel-model describes reduction as explanation (see below, Sect. 7.2), models of functional reduction often describe reduction as explanatory, and Melnyk (2003, Chapter 3) regards one form of reductive relation to be crucial for his version of physicalism – he argues that the form of physicalism he defends, which builds on the notion of multiple realization, is compatible with reductive explanation of, intuitively, higher level facts (this is not the term he uses on this occasion) in terms of physical facts and necessary truths (Melnyk 2003, 83).

discussed below (see Sect. 9.3). For the moment, suffice it to say that we adequately characterized one form of identity-based reduction in terms of a specific sort of explanation.

But can't we say more about this sort of explanation? Sometimes, philosophers have intuitively described reductive explanation as *mechanistic* explanation (Fodor 1974; Chalmers 1996; Kim 2008) – an idea also alluded to in the intuitive descriptions assembled in Chap. 2 (Wimsatt 2006). The remainder of the present Chapter is dedicated to a discussion of the relation between reduction, as captured by *(Explication I)*, and what has been intensively discussed under the title 'mechanistic explanation' within the past few years. Building on this discussion, the next chapter suggests that differences in meaning in reductive explanation reflect differences in *modes of presentation*, or differences in the way an object is presented *as having specific properties*.

Mechanistic explanation is usually described in terms of 'by'-explanations. Before turning to the details of the relation between reductive and mechanistic explanation, let me firstly argue that there is a variant of *(Explication I)* which comes in 'by'-format as well.

4.2 Reductive 'By'-Explanations

Craver and others deny that mechanistic explanation is reductive (see below). To tentatively motivate the idea that there is a tight connection between reductive and mechanistic explanation, consider candidates for reductive 'by'-explanations: 'things freeze *by* forming lattice structures', and 'things pump blood (naturally) *by* the rhythmic contraction of muscles in an appropriate environment.'[3] Here are the corresponding reduction statements:

(1) Freezing reduces to lattice-structure formation, and:
(2) (Natural) pumping of blood reduces to rhythmic muscle-contraction in an appropriate environment.

Even (1) and (2) express falsehoods, we can learn something about reduction pretending that they are true. And here are the truth conditions, now in terms of 'by'-explanations:

(1) Freezing reduces to lattice-structure formation iff

   (i) For every x, if x freezes, then x freezes *by* forming lattice structures, and
   (ii) Freezing = Lattice structure formation.

[3]There are numerous different formulations, instantiating different forms. Here are two others: "The heart pumps blood by *coordinated muscular contraction*" (Doerschuk et al. 1977, 258); "[T]he heart pumps blood by continuously and rhythmically contracting and relaxing" (Byrnes and Jensen 2001, 25). An interpretation of how these formulations interrelate would take us too far from our actual target. It would be interesting enough if at least one of them fits a reductive interpretation.

(2) Natural pumping of blood reduces to rhythmic muscle contraction in an appropriate environment iff

  (i) For every *x*, if *x* pumps blood (naturally) then *x* pumps blood by the rhythmic contraction of muscles in an appropriate environment, and
  (ii) the natural pumping of blood = rhythmic muscle contraction in an appropriate environment.

So, *prima facie*, some by-explanations figure in truth conditions for at least some reduction statements. We thus arrive at a characterization of the reduction of *events* (here, again, construed as types):

(Explication II – Event-Reduction) ϕ*-ing* reduces to ψ*-ing* iff

(i) for every *x*, if *x* ϕ*-s* then (*x* ϕ-s by ψ*-ing*), and
(ii) ϕ*-ing* = ψ*-ing*.

The importance of the distinction between event-reduction and kind-reduction should not be overestimated. It will be shown that reduction, which can be defined in terms of 'by'-explanations can also be defined in terms of 'because'-explanations. The idea is that if x ϕ*-s* by ψ*-ing* then x ϕ-s *because* x ψ*-s* (Schnieder 2009, Sect. 4.5 suggests that the relevant 'by'-statement implies a sentence of the form 'it is true that x ϕ-s *because* it is true that x ψ*-s*', Schnieder 2009, 666) and, building on this idea, if x ϕ-s *because* x ψ*-s* and, in addition, if (x's) ϕ*-ing* = (x's) ψ*-ing*, then x ϕ*-s* by ψ*-ing* (see below, Sect. 4.5.2). However, since in the debate on mechanisms, grammatical differences between 'because'- and 'by'-explanations have often been emphasized, it is worth noticing that at least some reductive explanations, and truth-conditions for reduction statements, can come in the 'by'-format. Thus, the claim that reductive explanation is mechanistic is not obviously mistaken.

4.3 Mechanistic Explanation – The Background

Is reductive explanation basically mechanistic in nature? To evaluate this point, we need some background on approaches to mechanistic explanation first. One common criticism of Nagel's model (and similar approaches to reduction) is that it is concerned with theories (and sciences) only (for an attempt to show that this criticism is misguided, see my (van Riel 2011b)). According to Wimsatt (1976), Darden and Maull (1977), and Schaffner (1993, 2006), this poses a problem especially for sciences that do not contain full-fledged theories, such as biology. A more general worry was that explanations, as construed by Nagel and what one might call the reductive tradition, interpreted explanations as instantiating *argument-schemata* – what was to be explained was regarded as being derivable from the material that served as (or expressed) the explanans. This *covering law* model of explanation (basically any variant of the DN-model) was rejected for several reasons we do not have to go into here (a summary can be found in Craver 2007, chapter 2). This sort of criticism inspired philosophers working on *mechanistic explanation* (Kauffman 1970; Cummins 1975, 1983, 2000; Bechtel and Richardson 1993; Bechtel 1994; Glennan 1996: Machamer et al. 2000; Craver 2005, 2007). Mechanistic explanation comes under many different labels. Sometimes it is described as *constitutive* or *structural*, and it seems to be related to *functional* explanation. The main idea is nicely captured in the following quote taken from Salmon (1984):

> A constitutive explanation is an exhibition of the internal causal structure of the explanandum; such an explanation exposes the causal mechanism within the explanandum. (Salmon 1984, 270)

Constitutive or mechanistic explanation explains something in terms of its constituents, or its underlying mechanism. But what is a constituent, or a mechanism of something?

In a first step, I will discuss the notion of a mechanism as conceived of within the debate on mechanistic explanation. In a second step, the role of specific question-answer pairs for the individuation of the relevant kind of explanations will be examined. Intuitively, these two aspects correspond to two different ways of characterizing an explanation – an *ontological* and a *grammatical* one. In a third step, I will point to some problems these candidate characterizations face, and a solution to these problems will be offered. Finally, a characterization of a kind of explanations will be given that covers reductive explanation as a special case.

Before entering the debate, let me briefly mention what I will *not* focus on, too: Epistemological and pragmatic issues of this kind of explanation will largely be ignored. A great deal of the work of philosophers who describe scientific endeavors as being related to mechanistic explanation consists in the investigation of actual scientific procedures. To this extent, these philosophers work with a paradigm that has been widely accepted in the Post-Nagelian tradition, and which has been described as being based on *extension first* approaches to reduction in Sects. 2.2 and 2.3. The target explicitly is the *description* of aspects of scientific investigation (Craver 2007, vii; Bechtel 2008, ix, though Craver assumes that his proposal has normative consequences as well).[4] These rich debates on actual examples especially

[4]It is worth noting that the mechanistic approach to explanations is associated with an anti-reductionist tendency in the philosophy of science. In their (1977) paper, Darden and Maull criticized the classical reductionist view, which can be regarded as a *mixtum compositum* of Nagelian reduction and the Oppenheim/Putnam picture of a hierarchy of sciences. They claim that "it becomes natural to view the unity of science, not as a hierarchical succession of reductions between theories, but rather as the bridging of fields by interfield theories" (Darden and Maull 1977, 61). The observation that in fact, explanations in high-level science are often multi-level (Wimsatt 1976; Darden and Maull 1977; Bechtel and Richardson 1993; Bechtel 1994; Schaffner 2006) and at least superficially distinct from explanations as classically conceived (by answering non-causally construed 'how'-questions as opposed to causally construed 'why'-questions) is perfectly compatible with believing the notion of reduction discussed here to be relevant: Even if general reductionism is not even a regulative idea of science, it might nevertheless be the case

in biology and cognitive science will be neglected. Thus, the structure I impose on the debate – the distinction between ontological and grammatical characterizations – has the status of a reconstruction. In the debate, these are intertwined with each other as well as with discussions of other aspects of mechanistic explanation.

4.3.1 The Ontological Characterization of Mechanistic Explanation

A mechanism is, according to one standard interpretation, a complex entity, consisting of constituents that are *organized to do something*.[5] An appropriate description of a mechanism explains the occurrence of the phenomenon to be explained. Characterizing mechanistic explanation in terms of what it is concerned with, i.e. in terms of how it relates to the worldly stuff whose interconnections give rise or correspond to mechanistic explanations, is characterizing it in *ontological* terms. It is similar to characterizing causal explanation giving an account of causal dependence, and it is similar to characterizing conceptual explanation in terms of the relations between concepts conceptual explanations allegedly track. This is an important (implicit) assumption: That on the one hand, we have explanations, whereas on the other, there are worldly relations corresponding to explanations, in virtue of which explanations are true or false. On this picture, a causal explanation is causal only insofar as it is (purportedly) about or corresponds to a causal relation, and it is true only if (and if so, in virtue of the fact that) the relevant causal link is instantiated out there. A conceptual explanation is conceptual only insofar as it is (purportedly) about or is concerned with a conceptual link, and it is true only if (and if so, in virtue of the fact that) the relevant conceptual link holds. Let us refer to such relations as 'explanatory relations'. They are explanatory in the sense that if they hold, there is at least one corresponding explanation that is true in virtue of the fact that this explanatory relation is instantiated. Consider the following explanation: He drank the water because he was thirsty. The explanation is a true causal explanation if and only if (and if so, in virtue of the fact that) his being thirsty, the event, caused his drinking the water, another event (the *target* of the explanation). We characterize an explanation referring to a corresponding relation that is explanatory in nature.[6]

---

that for some areas of investigation, parts of sciences or fragments of theories, reductive relations hold. This is also Schaffner's more modest point, that *partial* reductions are relevant (Schaffner 2006), see Sect. 7.6.

[5]For a comprehensive illustration of the differences between definitions of mechanisms, see (Bechtel and Abrahamsen 2005, footnote 5).

[6]Note that if you prefer a more ontic interpretation of explanations, you could, for the discussion to follow, assume that explanation-relations are themselves explanatory relations, and that what corresponds to explanatory relations are ways of presenting an explanation within language. This difference becomes apparent in the passage quoted above:

On this view, we can distinguish between kinds of explanations by appealing to corresponding explanatory relations. Conceptual explanations are different from causal explanations insofar as they correspond to different explanatory relations.[7] This is one of the two main strategies pursued by philosophers who work on mechanistic explanations. They describe mechanistic explanations in terms of the worldly relation purportedly corresponding to mechanistic explanations. Let us call this relation 'mechanistic dependence'. An explanation is mechanistic if and only if it is (supposed to be)[8] about or concerned with a mechanistic dependence relation.

For what follows, it will prove useful to mark a distinction between the *explanandum* (and an *explanans*) as a *proposition*, and the *entity* that figures as the *target* of an explanation. Explanans and explanandum are here conceived of as conceptual contents. Consider again the following example: He drank the water because he was thirsty. The explanans is the proposition that he was thirsty; the explanandum is the proposition that he drank the water. The target of the explanation

---

> A constitutive explanation is an exhibition of the internal causal structure of the explanandum; such an explanation exposes the causal mechanism within the explanandum (Salmon 1984, 270)

Here, the expression 'explanandum' is used in an ontic way. The *explanandum* is here interpreted as an entity (type) – an idea common among those working on mechanistic explanation. It is *things* and *events* that have some internal causal structure and, according to this picture, instantiate explanation relations.

[7]Note that this is not to say that expressions such as 'because' or 'by' used in explanations are ambiguous (this is an additional claim). Here is an argument to the effect that prima facie, we should not assume that even if different kinds of explanation correspond to different sorts of explanatory relations, 'because' is ambiguous. Assume that a crazy wizard brought it about that whenever Peter uses a match, a tree next to him burns down, even if there is no causal connection between Peter's using the match and the tree burning down, unbeknownst to Peter. Assume that similarly, Paul, dedicated environmental activist, does not know about this strange connection. Now assume that at some point, Paul sees Peter and the crazy wizard standing close to the still glowing ashes of a once beautiful redwood tree. Paul asks Peter what happened; Peter shrugs, and the wizard answers: 'Well, the tree burned down because Peter lit a match ...'. At this point, Paul interrupts him blaming Peter for what he did. After a while, the wizard manages to continue, '... but there was no causal connection between his using the match and the tree burning down – it was a miracle.' Did Paul misunderstand Peter in the first place? In a sense he did. He drew an incorrect conclusion about the connection between using the match and the tree burning down. But was this a kind of misunderstanding comparable to that of ordinary ambiguous expressions? It seems not. For example, it seems intuitively clear that Paul was in a position to know that the tree burned down because Peter lit the match, even if he misunderstood the situation. This is not the case for ordinary ambiguous expressions: If in 'Fa', 'F' is ambiguous, and Paul interprets 'F' such that 'Fa' is false, he is not in a position to know that Fa. Thus, the truth of 'The tree burned down because Peter lit the match' is independent of whether or not the relevant relation is causal or non-causal. This shows that, prima facie, 'because' is not ambiguous. The causal 'because' is, then, something like this: It is a relationally individuated 'because' – individuated with respect to (i) occurrences in sentences and (ii) these sentences relevantly corresponding to or being about causal relations.

[8]Why 'supposed to be'? Because explanations, as the term is used here, might be false.

is *his drinking the water* – an event. This is a verbal point; one may want to label the target 'explanandum'. Some such distinction is needed to reconstruct the relevant aspect of the discussion about mechanistic explanation, although it is not always clear what the target or object of an explanation is supposed to be. Intuitively, the target of an explanation is the entity the explanandum of an explanation is *about*. For some cases, it will be relatively easy to identify the target of an explanation. In an explanation of why a specific Tsunami occurred, the target is the occurrence of a specific Tsunami. Asking how humans are able to cognize the actions of others as intentional, the target is the human ability to cognize the actions of others as intentional. Sometimes, however, things are more complicated. What is, for example, the target of an explanation of why salt dissolves in water? Is it the disposition of salt to dissolve in water? Or is it an event type, i.e. the event of salt dissolving in water? Or is the target just the proposition or the fact that salt dissolves in water? These questions may not pose insurmountable problems; but trying to answer them would take us too far from our actual target. Let us rely on an intuitive understanding of the notion of a target of an explanation – we will need it for heuristic purposes only.

On the ontological interpretation of what it is for an explanation to be mechanistic, we have to give an account of mechanistic dependence. A mechanism is what is supposed to do the explanatory work in a mechanistic explanation; intuitively, a mechanism is what gives rise to the target. There are a number of different characterizations of what a mechanism is. I will work with a minimal characterization. First of all, it is important to note that mechanisms are mechanisms *of some phenomenon* (Glennan 1996, 2002; Craver 2001, 2007) to be explained by that mechanism. This motivates tying the notion of a mechanism to what it does, an idea reflected in Craver's characterization:

> Mechanisms are collections of entities and activities organized together to do something. (Craver 2005, 385)

On most accounts, the notion of a mechanism depends on the notion of *organization* (cf. Bechtel and Richardson 1993; Glennan 1996, 2002; Machamer et al. 2000; Bechtel and Abrahamsen 2005; Bechtel 2007). This is what gives, on these accounts, rise to the explanatory link. The explanatory relation is instantiated by the target and what gives rise to the target. The heart pumps blood by the orchestrated interaction of its constituents. The constituents are organized to give rise to the heart's pumping blood.

For what follows, we should rely on an intuitive understanding of organization (we do not get more than that in the debate on mechanistic explanation): Computers and their functioning, perceptual systems, cells, and bodily parts are organized to do something.[9] For mechanistic explanation, we can then rely on Cravers description,

---

[9]One common way to approximate the notion of organization (or a mechanism) is to contrast it with Wimsatt's notion of an aggregate (Bechtel 2007; Craver 2007; for the original idea, see Wimsatt 1997, 2006). An aggregate of an object x is a property of x with respect to specific decompositions of x into parts (and their properties), which, intuitively, remains unchanged (or changed only in

giving a sketch of the notion of a mechanism (the notion of an entity is here used in a wide sense, such that we do not have to mention activities in addition): *Something is a mechanism iff it is a collection of entities, which are organized to do something.*[10] And something is a mechanistic explanation iff it corresponds to the instantiation of a mechanistic dependence relation, where 'correspondence' just mimics the intuitive 'being (purportedly) about'.

The characterization of mechanistic dependence is, however, still rather meager. We may hope to learn more considering the notion of a component in or of a mechanism. As Craver puts it:

> Mechanistic explanations are *constitutive* or componential explanations: they explain the behavior of the mechanism as a whole in terms of the organized activities and interactions of its *components*. (Craver 2007, 128)

We should take this literally: Reference to capacities of components is required to make an explanation mechanistic. So, what is a component of a mechanism?

Components of mechanisms are supposed to be mechanisms themselves (with the possible exception of components at some fundamental level). To avoid long exegetical remarks, I suggest the following interpretation: A *mechanism*, an ordinary object or entity, is a collection of other ordinary entities,[11] which, together, engage in an event that *explains* the event produced by the mechanism or by another object the mechanism belongs to. Not every proper part of the object that produces the phenomenon to be explained needs to be a component in the mechanism. The entire mechanism may be a proper part of *another* object that performs a certain task that depends on the functioning of the mechanism. Consider a watch and its mechanism.

---

value) under decomposition and reaggregation of x's parts, under substitution of equivalent parts and under rearrangement of parts of x, under adding parts to x, and which does not depend upon any causal interaction between x's parts (Wimsatt 2006). Relying on an intuitive notion of organization will prove useful here, since the notion of organization (being a teleological notion) will be changed later on; reduction is not only concerned with objects being the possibly subject of a teleological description.

[10]It is noteworthy that the notion of a mechanism seems to convey some naturalist co-notation, which is, in this definition, not explicitly covered: intuitively, mechanisms are opposed to souls and spirits. This would limit possible applications of reductive explanation as mechanistic explanation to possible cases of reduction to spiritual things – an option our description of reduction should leave room for.

[11]Some remarks on components are rather confused. Craver (2007) for example argues that mechanisms and components are not ordinary objects, and that our usual ways of classifying things into events, properties, substances and the like falls short of an appropriate characterization of the metaphysics of mechanisms and their components. Craver discusses this issue in the context of mechanistic levels. He assumes that these levels of mechanisms are not levels of objects, like 'societies, organisms, cells, molecules and atoms' (Craver 2007, 190). Rather, they are levels of 'behaving components' (Craver 2007, 190). Thus, components (which are themselves mechanisms) are not objects, but rather 'behaving components'. Roughly, the confusion seems to be due to a misunderstanding: Some of the expressions used to designate components in a mechanism designate events; blending this with mereological talk about proper parts leads to the confusion that there are what is sometimes called "behaving components" of a mechanism, which are not to be conflated with ordinary objects.

A watch represents times successively. Representing these times successively is an event. A watch contains proper parts. These objects – gears, screws etc. – engage in doing something, they are bearers of events, the occurrence of which brings about the functioning of the watch. Some parts of an ordinary watch belong to its mechanism, whereas others do not. Assume that one of the gears in our watch bears the watchmaker's signature – a physical object consisting of, say, ink. This may be a proper part of the watch, but it does not contribute to its functioning. In this case, the mechanism is a proper part of the watch. Other objects may have a mechanism as an improper part. In that case, the object is identical to the mechanism. There could be a watch that is identical to its mechanism, representing times successively by the mechanism being in different states. The functioning of both kinds of watches mechanistically depends upon the events its mechanism's components produce.

Often, it is suggested that the relation of *being a component* is non-causal (Bechtel 2007; Craver 2007): The functioning of the constituents of V1 (an area processing visual information) does not *cause* V1 to process information – these constituents and their functioning *constitute* V1's processing information. Thus, we have a relation between a complex, encompassing event, the processing of information by V1, and the events underlying or constituting this event. Giving an appropriate description of the mechanism in terms of its components or constituents, we should mention the causal organization of the mechanism's constituents (Craver 2007, 62, this also underlies Salmon's 1984 and, in a similar fashion, Causey's 1977 conception of constitutive explanation), because it is their causal functioning that explains the functioning of the whole. Thus, it is assumed that it is the component's causal organization that gives rise to the mechanism's behavior. This idea of *giving rise* to something – an explanatory notion, as the quote from Craver suggests – is crucial: It is the notion corresponding to mechanistic dependence we need in order to come up with an appropriate characterization of mechanistic explanation (on the ontological conception). The idea seems to be this: *A part x of an object that has mechanism M to produce E is a constituent or component of M if and only if x produces some event y, such that M's producing E occurs (partly) in virtue of the occurrence of y.*

Note that here, we make use of the notion of 'in virtue of' in order to define the notion of a constituent, which, in turn, enters the characterization of a mechanistic explanation; this is somewhat similar to the way in which 'reduction' was defined in *(Explication I)* above. This is obviously not to say that this isn't an interesting definition of what a component in a mechanism is supposed to be. However, following this line of thought, we must not hope to come up with a "reductive" definition of mechanistic explanation, unless we possess a general strategy to get rid of 'in virtue of'- and other explanatory talk.

To sum up: On the ontological account of mechanistic explanation, an explanation is mechanistic iff it corresponds to the instantiation of a mechanistic dependence relation. Correspondence is still an intuitive notion: Just like the explanation that he drank water because he was thirsty is causal in virtue of somehow corresponding to an alleged causal relation, mechanistic explanations are supposed to correspond to mechanistic dependence relations.

The *relata* of a mechanistic explanation are events – one event the mechanism (or another object) is the bearer or the subject of, and an event its components are the bearers or the subjects of. To clarify this jargon: Persons are the "bearers" of their actions, houses are the "bearers" of their burnings, and watches are the "bearers" of their representing times. In the light of this, we can characterize mechanistic dependence: one event mechanistically depends on another event if and only if the latter is an event the bearer of which is a component in the mechanism to produce the former. And when is an object a component in a mechanism? We already gave the sketch of an answer above. To keep things simple, let us focus on the case where the mechanism is identical to the object that produces the target phenomenon. Here is the characterization again, in a more precise form (schematically):

(Components of Mechanisms): x is a component in the *mechanism, y,* to ϕ *iff y* ϕ*-s (partly) by* the ψ*-ing* of x.

One may want to add a nomological necessity operator to the right hand side, so as to ensure that no arbitrary, merely temporal dependence relation is covered. In the context of the life sciences, we might hope to substitute this talk by 'being selected for'-talk; and in the case of artifacts, such as watches, we might want to limit this talk to talk about intended use, or *de-facto* use within a social group. Thus, one may add a conjunct that x was *selected,* or *chosen* for ψ*-ing*. For present purposes, this characterization is sufficient. It yields an ontological characterization of mechanistic explanation in the sense that it enables us to characterize mechanistic explanation in terms of a corresponding ontological relation of mechanistic dependence, which, in turn, rests on the notion of a component in a mechanism. There is, however, another strategy in the debate: That of characterizing mechanistic explanation in terms of certain grammatical surface-properties of sentences that express these explanations.

4.3.2 The Grammatical Characterization of Mechanistic Explanation

In an ontological characterization, kinds of explanations are distinguished in terms of corresponding ontological relations. On what I have labeled 'grammatical accounts', explanations are distinguished in terms of different linguistic features exhibited by different explanations.

Following Craver and Bechtel, mechanistic explanations are constituted by answers to certain 'how'-questions (in a non-evaluative sense, as one should hasten to add (cf. Schnieder 2009)). Such questions take the form 'How does x φ?' A corresponding explanation is, then, a 'by'-explanation of the following form '*a* φ-s by (*b*'s) ψ-ing'. Reference to 'b' is relevant insofar as we sometimes face explanations such as 'the heart pumps blood by *the organized contraction of its constituent muscles*'. To connect this to the less precise talk above: In the position of '*a*' or '*b*' we substitute an expression that designates the bearer of an event; in the case of '*a*', it is the mechanism, in the case of '*b*' it is the component, or the collection of components. Note again that this characterization perfectly matches the idea that what is to be explained is the occurrence of an event.

But does this grammatical characterization suffice? It does not, as long as we intend to capture *mechanistic* explanation *only*. Two possibly problematic cases have been recognized in the debate: that explanations that refer to mere models as well as non-mechanistic functional explanations may come in the same format. Therefore, the grammatical description is flanked by additional requirements:

Firstly, mechanistic explanations correspond to, as Craver has recently pointed out, pairs of 'how-*actually*'-questions and their answers. Intuitively, a 'how-actually' question is to be distinguished from a 'how-possibly' question in that an answer to a 'how-actually' question has to refer to the actual mechanism which *in fact* produces or constitutes the phenomenon to be explained (Craver 2007, Chapter 4 – here, one also finds a detailed discussion of the distinction between these two kinds of questions). *Mere models* in science give answers to 'how-*possibly*' questions. Now, Craver conclusively argues that it is this aspect that increases the explanatory power of mechanistic explanations (Craver 2006, 2007, chapter 4). Following Robert Cummins (1975, 1983), Daniel Dennett (1994), and Herbert Simon (1969), Craver argues that giving descriptions in terms of homunculi, subsystems, reverse engineering, decomposition or functional organization is a useful guide to the relevant sort of constitutive explanation. He argues that these strategies have to be accounted for in terms of aiming at explanations that reveal what actually happens.

Here is an interesting side effect: Knowledge of answers to 'how-*actually*' questions will increase the *manipulability* of the phenomenon we are interested in (it gives, intuitively speaking, a more detailed list of possible nodes we can influence) (Craver 2006). This is an interesting epistemological and pragmatic point: If we were able to give a detailed description of the causal organization of a phenomenon, we would be in the position to give a more detailed list of possible interventions (Woodward 2003; Pearl 2000). Based on mere models, this will be difficult: The objects postulated by a mere model just do not exist. An *ideal* mechanistic description would give us all the relevant aspects of mechanistic organization (Craver 2006). Mechanistic descriptions we are able to give do less than that, but they nevertheless increase our understanding of how things work and, thus, increase the ways in which we can intentionally intervene on them – a topic picked up again in Sect. 9.4.

Secondly, asking how Oswald killed Kennedy, we may seek an answer in terms of the actions he performed that constituted his killing Kennedy. An appropriate answer to this question is this: He killed Kennedy by shooting him.[12] This is an

[12] Philosophers have taken different stances towards the question of how actions should be individuated. Some hold that in our case, there is one action under different descriptions (Anscombe 1957; Davidson 1969), others assume that there is more than one action (Kim 1976; Goldman 1970).

instance of a 'how-actually' explanation, but it is not a mechanistic explanation. Rather, it is what is sometimes referred to as a *functional* explanation. In another context, this has been recognized by Craver (2007, 129), who takes this point from Cummins (1983). I suggest that nevertheless, constitutive explanation should be regarded as a variant of functional explanation. Recall: it is the constituents' causal functioning that explains the functioning of the whole. The moral to be drawn from these points is rather important: characterizing mechanistic explanations as corresponding to pairs of 'how'-questions and their answers does *not* fully capture the notion of mechanistic explanation. Moreover, as already indicated in Sect. 4.2, it will turn out that there is no sharp distinction between 'by'-explanations and 'because'-explanations (see Sect. 4.5.2). If there are no further grammatical differences available, an appropriate characterization of mechanistic explanation has to build on ontological criteria, say, concerning the difference between constituents on the one hand, and what is reflected in the debate on action-individuation as the accordion effect on the other.[13] As will become apparent below, there are further difficulties with this characterization: First of all, 'how' questions do not play the prominent role they purportedly play in the context of mechanistic explanation, and a family of explanations, namely, reductive explanations, transcends the boundaries determined by the focus on *organization* and the rather demanding notion of *components*.

This notion of mechanistic explanation in mind, let us turn back to the question of how mechanistic explanation relates to reductive explanation. Based on the rough characterizations just given, it will now be argued that mechanistic explanation perfectly matches the idea that we get diversity in terms of differences in conceptual material, unity in terms of identity, and directionality in terms of explanatory dependence. In order to see how unity relates to mechanistic explanation, let us first consider the notion of mechanistic levels – it forms the guide to understanding unity in models of mechanistic dependence.

4.4 Identity in Mechanistic Explanation

Is reductive explanation mechanistic explanation? If so, the conception of mechanistic levels ought to be compatible with strong unity in a mechanistic hierarchy. The following bi-conditional states the suggested connection between level-talk and the property of being a component of a mechanism:

> X's φ-ing is at a lower mechanistic level than S's ψ-ing if and only if X's φ-ing is a component in the mechanism for S's ψ-ing. (Craver 2007, 189, similarly in Craver 2005, 389 and in his 2007, 170)

[13] Searle (2010) ascribes the invention of this term to Feinberg (1970, 34).

Levels of mechanisms are thus to be understood as levels of components which, in turn, can be accounted for in terms of their components, which, in turn, can be accounted for in terms of their components, and so on.[14] Based on the relation of *being a component in*, we get the following picture of hierarchies of mechanistic levels: Let us call the mechanism we want to explain on a given occasion the *target-mechanism* (just like in general, what we want to account for in an explanation the *target* of that explanation). A component in one mechanism may be the target of another mechanistic explanation.[15] A mechanism's components are organized in a hierarchical structure: Some constituents of the target mechanism are themselves constituted by other mechanisms, which in turn are constituted by other mechanisms and so forth. The constitution-relation gives an idea of a hierarchical organization of levels of mechanisms. It also matches the idea of explanation: We explain a complex

[14]The term 'level' is notoriously ambiguous. One common way to conceive of distinctions between sciences and scientific levels (in the ordinary sense of the expression 'science') is in terms of some paradigm or shared assumption about how to account for a set of problems. Let me briefly argue that these aspects do not give rise to an interesting sort of reductionistic hierarchy and, thus, do not help distinguishing between the relevant levels, even though they might be of utmost importance for what makes a science (an object comprising psychological, social and pragmatic aspects) the science it is. Wimsatt, for example, gives a detailed account of scientific levels, characterizing his prototype-account (Wimsatt 1976). In addition to claiming that levels are partly individuated by objects (here: objects of similar size) and relations among these objects, Wimsatt refers to techniques of investigation and disciplinary perspectives that play a crucial role in the individuation of scientific levels. One could easily see that these latter features are irrelevant for a notion of a scientific level that supposedly corresponds to the notion of reduction. Change in disciplinary perspective and change in techniques of investigation by preservation of some set of objects and kinds does not necessitate reduction, even though it will lead to some sort of theory succession. Similar to Wimsatt's idea of perspectives, Darden and Maull (1977), Kuhn (1962), and Lakatos (1977) account for sciences partly in terms of what is the socio-psychological or epistemological glue holding different theories together. Craver describes this glue as amounting to level-distinctions (Craver 2007, 171). He seems to assume that it makes sense to talk about different levels as being solely distinguished by different paradigms (or similar aspects). Maybe it does. However, it should be obvious that this talk does not make for reductive levels – shift in paradigms does not necessarily make for straightforward reductions, but allows for replacements. For a purely ontological conception of levels, see Kim (2002).

[15]According to Craver (2007, 190) and Bechtel (2007, 550), mechanistic levels are not supposed to be 'monolithic divisions in the structure of the world' (Craver 2007, 190). Thus, it is not the case that for any two kinds of components, one is a component within the same hierarchy as the other. At first sight, this does not come as a surprise. However, classical conceptions of scientific hierarchies tend to suggest that level-talk commits us to the assumption that there is one unique hierarchy where every scientifically relevant kind finds its place. This is the reading the Oppenheim-Putnam division of sciences (Oppenheim and Putnam 1958) has been given, and on that interpretation, it was criticized, for example, by Wimsatt (1976). The most important argument is epistemological in nature: Craver argues that we cannot 'read off' the mechanisms from 'a menu of levels in advance'. Rather, we have to answer questions of how many levels there are 'on a case-by-case basis by discovering which components at which size scale are explanatorily relevant for a given phenomenon' (Craver 2007, 191). Epistemological considerations aside: It should be obvious that a mechanistic hierarchy does not necessarily comprise all sorts of objects there are in the world, just because not all such sorts of objects are mechanistically related.

in terms of the arrangements and interactions of its parts (and their properties). Building on this picture, we may now turn to the question of whether or not mechanistic explanation is compatible with reduction, as conceived of here: Is it (conceptually) possible that the relevant object an explanandum is concerned with is identical to the object relevantly referred to in the explanans? Is it the case that necessarily, if S φ-s by the ψ-ing of some of its constituents, then S's φ-ing $\neq$ S's constituents' ψ-ing?

Craver (2007) shifts between reductionist and anti-reductionist descriptions of mechanistic (here figuring as 'constitutive') explanations. On one occasion, he writes:

> There are two dominant and broad traditions of thought about constitutive explanation: the reductive tradition and the systems tradition. My view is a development and elaboration of one strand on the systems tradition. (Craver 2007, 108)

Thus, "his view" opposes reductive interpretations of mechanistic explanation. Some pages later, however, the following intuitive characterization of the system approach is given: It is 'to reduce a capacity to the programmed exercise of sub-capacities' (Craver 2007, 110). So, is mechanistic explanation reductive, or isn't it? One might suggest solving this tension in the spirit of Kim (2008) and others (Chalmers 1996, 43, Fodor 1974, 107) who give an idea of reductive explanation as opposed to (identity-based) reduction in terms of mechanistic explanation. According to this view we can have reductive explanation *without* reduction as being based on identity. Kim goes a step further, claiming that we cannot have both at the same time: identity-based reduction and reductive explanation (conceived of as closely tied to mechanistic explanation) contradict each other. This would nicely fit Craver's way of talking – that mechanistic explanations do not yield reductions, but, nevertheless, reductively explain some capacity.

4.4.1 *Kim's Worry*

This is Kim's worry: Given that $a = b$, and a reduces to b, and a is to be explained in terms of b, we face a problem: Since nothing can be explained by itself (Kim 2008, 102 f., 106), we cannot have both at once, identity-based reduction and the corresponding reductive explanation. Put differently: Reductive explanation would essentially be trivial; any such explanation would be plainly false. Here is, however, a rival intuition:

> What reduction needs [...] is the idea that the 'reduced phenomenon' is made more comprehensible or intelligible by being shown to be identical with the 'reducing phenomenon'. (Crane 2001, 54)

If, say, water is made more intelligible by being shown to be identical to H_2O, then it seems that water (or its behavior) is thereby somehow explained. But how can an explanation foster our understanding and, at the same time, be trivial? This is impossible. So, who is right?

Kim's argument rests upon a misunderstanding. An explanation is trivial only if its explanandum is relevantly contained in its explanans. One straightforward kind of this containment is this: Explanations of the form '*p* because *p*' are trivial and false (if '*p*' is not relevantly ambiguous and disambiguated on both occasions). There might be more subtle kinds of problematic containment of the (parts of the) explanandum in the explanans. However, brief reflection upon what the relevant material, that is contained in the explanandum, *is* will help solve this problem: The connection between triviality of an explanation and containment of fragments of the explanandum in the explanans concerns, at best, the conceptual level, i.e. ways in which things are presented by the meaning of an expression.[16] It does not concern the 'metaphysical level', i.e. what is represented by the meaning or the conceptual content.

4.4.2 Kim's Mistake

What Kim misses is this: Explanations are not individuated by the properties and individuals their constituents pick out in order to explain, but rather by the way they present us with these properties and objects. This is pointed out by Ned Block (who uses the term 'fact' in the way I would use 'proposition'):

> Just as knowledge of the fact that freezing happened is not knowledge of the fact that lattice-formation happened, so also an explanation of the fact that freezing happened is not an explanation of the fact that lattice-formation happened. By contrast: just as the time at which freezing happened is the time at which lattice-formation happened, so the cause of freezing is also the cause of lattice formation. (Block forthcoming)

Block's point is that we can give different explanations which both refer to the same objects (cause) in the *explanans*. He argues that the cause of freezing = the cause of lattice formation. However, the explanation of why lattice-formation happened is different from the explanation of why freezing happened. This is so because the *explanans* and the *explanandum* have, in both cases, *different meanings*, although they give access to the same state of affairs, or law-like connection. For an explanation to be *trivial*, meaning identity, or identity of conceptual content, or identity of Fregean sense is required. This is Block's point, or so it seems: knowing that freezing happened is not knowing that lattice-formation happened, because 'lattice formation' and 'freezing' have different meanings. So, the principle that *nothing* explains *itself* seems to require quantification over entities *as presented by a certain meaning.* It should be clear that conceptual difference between *explanans*

[16]Maybe, it does concern the level of expressions only; if the explanans and the explanandum in a conceptual explanations, such as 'Donald is a drake because Donald is a male duck', are synonymous, and if synonymity is sufficient for conceptual identity, then even conceptual containment does not amount to triviality.

and *explanandum* is a *requirement* for the truth of explanations such as the explanation of the human capacity to cognize the other as social in terms of mirror neuron mechanisms. Similarly, nothing is water because it is water, though, in the spirit of (*Explication I*) one may want to suggest that water is water *because* it is H_2O. Thus, conceptual difference is, again, the relevant point here: Even though water is identical to H_2O, H_2O can, in some sense, "explain water". *H_2O* comes in the appropriate conceptual shape. Thus, difference in meaning is a pre-requisite for reductive explanation. But conceptual difference does not necessarily translate into ontological difference. The principle Kim seems to allude to, that nothing explains itself, is, as he interprets it, misguided.

4.4.3 Identity

There is, at least *prima facie*, no reason to assume that reductive explanation is incompatible with identity-based reduction. It is at least conceptually possible that the freezing of water just is the formation of lattice structures by H_2O-molecules. Thus, one candidate reason for denying the possibility of cross-level identities in explanatory hierarchies has been dismissed. Even if nothing can be explained by itself (no explanans should be identical to its explanandum), an explanans and an explanandum may still refer to or be about the same object.[17] But isn't there positive reason to assume that mechanistic explanation allows for the relevant cross-level identities?

S's *φ-ing* is explained in terms of an underlying mechanism, which consists of entities that engage in certain events, which, together, give rise to *S*'s *φ-ing*. '*S*'s *φ-ing*' is an expression that designates an event in a mechanistic explanation. A mechanistic explanation is an explanation of an event-type, some *S*'s *φ-ing*, *which is explained by interactions of components of S*. Recall the idea of a component: A component is the bearer of an event in virtue of which the relevant activity of the overall-mechanism occurs. Components are parts of mechanisms. But what is the relevant relation between their orchestrated interaction and the event with respect to which they count as components of the mechanism (unlike other parts of the object the mechanism is a mechanism of, like the signature that belongs to the watch, but does not belong to its mechanism)? Here is a thesis: *At least for some mechanistic explanations, the event described in the explanandum is identical to the event described in the explanans.* And here are three arguments:

[17]In his (2006) Schaffner suggests that something like mechanistic explanation may go together with identity. His rich description of the epistemic and pragmatic aspects does not involve an evaluation of how this relates to 'by'-explanations, and how alleged problems of such identifications could be solved.

4.4.3.1 The First Argument

The first argument is based on the observation that for some explanations, which are not paradigmatically mechanistic, this idea makes perfect sense. Assume that the freezing of water is nothing but the lattice-structure formation of sums of H_2O molecules. Now, how does water freeze? It does so by the formation of lattice structures of H_2O molecules. Assume that pain is identical to C-fiber stimulation. How do pains occur? They do by the occurrence of C-fiber stimulations. This matches one of the more intuitive examples for reductive explanation. Nagel suggests that when:

> ... the detailed physical, chemical, and physiological conditions for the occurrence of headaches are ascertained [...] an explanation will have been found for the occurrence of headaches. (Nagel 1961, 366)[18]

Similarly, assume that our cognizing others as social beings is nothing but the activation of mirror-neurons in motor-regions of our brains. How does the relevant mechanism enable us to cognize others as social beings? It does so by the activation of mirror-neurons in motor regions of our brains. And at least *prima facie*, there is no reason to assume that this does not generalize to other cases: The mechanism by way of which a watch represents times successively is nothing but the sum of the relevant components, and the event by way of which a clock represents times is nothing but the concerted activation of these components. First of all, it is conceptually possible. Secondly, from the point of the reductionist, it seems economical: Why should we postulate additional events when we have perfectly analogous cases that *suggest* the economical interpretation? In our paradigm cases, there is no reason to postulate a difference between the events described in the explanans and the explanandum. Why should we postulate a difference, without further ado, in the present case? Thus, we shift the burden of proof. One argument against the compatibility of mechanistic explanation and cross-level identities has been dismissed. Secondly, there are cases that suggest the identity-interpretation.

4.4.3.2 The Second Argument

The second argument supports the idea that at least sometimes, the identity-interpretation is correct. The Davidson-Anscombe thesis (see, for example, Davidson 1963, 1967; Anscombe 1957) states that *if x φ-s by ψ-ing, then x's*

---

[18] A structurally similar case is described in Nagel (1961, 434), where Nagel argues that we are not in a position right now to reduce biology to physics or chemistry. This goal would be achieved if we were able to formulate conditions in physical or chemical language for the occurrence of biological phenomena. This matches a prominent idea to be found, for example, in Friedman (1982, 17), who describes theory-reduction in terms of explanation of the phenomena of the reduced theory by the reducing theory. This idea already shows in Nagel (1935), where reduction is described in terms of constitution.

φ-ing = *x's ψ-ing*. According to this view, someone's pushing a button could be identical to her causing the death of a million people (namely, if that person killed a million people by pushing the button), my driving a car is identical to (a part of) my contribution to the pollution of earth, and so forth. This line of thought has, however, recently been disputed by Schnieder (2009). Roughly, Schnieder argues that instances of '*x φ-s by ψ–ing*' could be true, even if a corresponding instance of '*x's ϕ –ing* = *x's ψ–ing*' (which is implied by the Davidson-Anscombe thesis) is false. The idea is that for a 'by'-explanation to be true, the event relevantly related to the explanans could be a mere constituent, or a "part" of the event relevantly related to the explanandum. Identity is not required. For the present context, suffice it to say that nevertheless, there is reason to assume that at least sometimes, or possibly a sentence of the form '*x φs by ψ-ing, and x's φ-ing* = *x's ψ–ing*' expresses a truth. This would be sufficient to establish the point we are concerned with here. The generality of the Davidson-Anscombe thesis might be mistaken; this does not imply that necessarily, it is never the case that this form of explanatory dependence goes together with identity, at least under a reading of events as particulars.

4.4.3.3 The Third Argument

I think that one of the intuitions that play a major role in the reduction debate is this: We can sensibly ask questions such as 'What *is* (the nature of) water?' Or: 'What *is* (the nature of) the mechanism that keeps the watch running?' Or: 'What *is* (the nature of) the mechanism that enables us to cognize others as social?' Sometimes, questions of this sort seek a conceptual explanation. But they may also aim at an answer that reveals the nature of an object, a property, a fact, or an event. If naturalism is true, then answers in terms of molecule-structure, in terms of the interaction of gears and screws, of mirror neurons and synchronization of synaptic activity are promising candidates for appropriate answers. And they may be accompanied by a 'nothing but'. Water is, accordingly, *nothing but* H_2O. Hence, it is identical to H_2O. The mechanism that keeps the watch running is nothing but the structured object consisting of gears, screws and the like. And, although the content of this claim may be disputable, some have claimed that the mechanism that enables us to cognize others as intentional agents is nothing but the mirror-neuron mechanism. On a weak interpretation, 'by'-explanations are at least compatible with corresponding 'nothing but' statements. Hence, they are not incompatible with the relevant identity-links. On a more thorough interpretation, it is the search of what these things really, or ultimately are, what their nature is, that gives rise to or justifies the relevant by-explanations. Both interpretations suit the point to be established here: 'by'-explanations are compatible with the relevant cross-level identities.

Thus, it seems perfectly sound to assume that at least some mechanistic explanations are such that they reconcile conceptual diversity with strong unity. Even if reductive explanation is not basically mechanistic, it is at least a cognate of mechanistic explanation in the sense that mechanistic and reductive explanation overlap. In a mechanistic reductive explanation, the relevant directionality stems from an explanatory relation. What has been proposed as a reductive 'by'-explanation in Sect. 4.1 seems to be a cognate of mechanistic explanations. Nevertheless, there are important differences. First of all, reductive explanation does not hinge upon the idea that things are *organized to do* something – this teleological aspect is irrelevant to the broader notion of reductive explanation. Some obviously non-mechanistic explanations, explanations that do not involve reference to organized things, such as the explanation of the occurrences of the freezing of water, are reductive. Moreover, the focus on 'how'-questions and 'by'-explanations does not play the role in reductive explanation it is supposed to play in mechanistic explanation according to the model presented above. Emphasizing 'how'-explanations seems to be a result of the focus on events in theorizing about mechanistic explanation. Reductive explanation should be construed more liberally – that a dollar bill is money may reduce to its being accepted as money, water reduces to H_2O, and the fact that this particular amount of water freezes reduces to the fact that this particular sum of H_2O molecules forms lattice structures. In these cases, no events are alluded to. Let us address these points, before applying some of the lessons drawn from a discussion of mechanistic explanation to a more detailed model of reductive explanation in the next chapter.

4.5 Mechanistic and Reductive Explanation

We now consider the model of mechanistic explanation as a candidate for a model of reductive explanation of events in general. Does mechanistic explanation, if appropriately combined with identity, make for reductive explanation? Identifying aspects in the characterization of mechanistic explanation that distinguish it from reductive explanation enable us to fully characterize the relation between reductive and mechanistic explanation.

4.5.1 Teleology

Relying on the characterization given above, a mechanism is necessarily *organized to do something*. If the freezing of water reduces to lattice-structure formation of H_2O-molecules, and if the notion of a reductive explanation hinges upon the notion of a mechanism, then water-molecules are *organized to form* lattice structures. This is absurd. Thus, we should get rid of teleology.[19] I suggest replacing the notion of a

[19]Removing teleological talk, this model of explanation is basically to be found in (Salmon 1984). Salmon, however, focuses on the relation between aeteological explanation and constitutive explanation, basically arguing that to account for causal explanation, we have to account for both

mechanism by the notion of a *constitutive-structure*.[20] The notion of a constitutive structure is supposed to *cover* that of a mechanism. Thus, the mechanism of a watch is, with respect to its functioning, its constitutive structure. The mechanism of human social cognition is a constitutive structure with respect to events of social cognition, namely the structure that constitutes events that are instances of social cognition. But whatever the freezing of water depends upon, it is not a mechanism. Water just freezes without being organized to freeze. It freezes because it is *constituted* in a certain way, without being organized to do something in a way. In a constitutive explanation, we should describe an object in terms of its constituents and in terms of these constituents' properties. However, a constituent is not necessarily a component in a mechanism.

The thesis is, then, this: *Reductive explanation* is constitutive explanation plus identity. If identity comes with constitutive explanation for free (a question not to be pursued here), then reductive explanation just is constitutive explanation (with the contention that partial reductions, as defined below, do not amount to, but can be defined in terms of identity-based reduction). But how does constitutive explanation relate to the 'grammatical' aspect of theories of mechanistic explanation?

4.5.2 Grammar

It has been assumed that mechanistic explanations that are constituted by answers to how-questions do not translate into 'because'- statements. Philosophers sometimes claim this to be a virtue of their mechanistic accounts (see, for example, Craver 2007): They argue that classical approaches to explanations like the DN-model, or unificatory models (one versions of which, namely Kitcher's 1981, 1989, is discussed and – rightly – dismissed in Craver 2007, chapter 2), miss the point of a great deal of explanations given in science.[21] Let me briefly point to a problem here. It seems that for a question of the form: 'How does *a* φ', that requires an answer of the form '*a* φ-s by ψ*-ing*', there are corresponding questions that come

these aspects. Whereas causal chains are described in terms of transmission of marks (intuitively: changes in character), the constitutive aspect gains little attention.

[20]Notions of *structural explanation* are well known in Philosophy. The term is highly ambiguous. When used in the context of constitutive explanation, as it is done here, it is often conceived of as some sort of causal explanation (McMullin 1978). There are at least two additional uses: The first is associated with the semantic program in the Philosophy of Science. Here, the term 'structural explanation', as used, for example, by Sneed (1994) picks out some sort of explanation based on a model-theoretic account of theories. Jackson and Pettit, on the other hand, use 'structural explanation' as it is often used in the social sciences, namely as some sort of macro-level explanation which, intuitively, *abstracts from the details* (Pettit and Jackson 1992). The sort of explanation focused on here is not to be conflated with these sorts of explanation.

[21]Recall Nagel's headache-example given above: obviously, these explanations did not escape his attention.

in the 'why'-format and seek a 'because'-explanation; or, as Schnieder put it, if '[t]hat x ψ-s explains how x φ-s', then '[i]t is true that x φ-s because it is true that x ψ-s' (Schnieder 2009, 666). Here are two candidates, and the corresponding explanations:

(i) 'Why does a φ?'
'a φ -s because a ψ *-s*'
(ii) 'Why did *a's* φ-ing occur?'
'*a*'s φ *–ing* occurred because *a's* ψ-ing occurred.'

Consider the question 'Why is this a water-constituent?'[22] The answer is this: 'This is a water-constituent *because* it consists of two hydrogen molecules and one oxygen molecule.' In response the question 'How did the heart pump blood?' we can say that 'the heart pumped blood *by* expanding and contracting regularly'. A similar explanation, now using the connective 'because', can be given in response to 'why'-questions: 'The heart pumped blood *because* it expanded and contracted regularly', and, less intuitive but, I submit, equally true: 'the heart's pumping blood occurred *because* its expanding and contracting regularly occurred'. Thus, reductive 'by'-explanations are covered by the explication given above:

(Explication I – Kind Reduction) *F-ness* reduces to *G-ness* iff

(i) for every *x*, if *x* is *F* then (*x* is *F because x* is *G*), and
(ii) *F-ness* = *G-ness*.

We can easily transform the explication reserved for reductive 'by'-explanations, namely *(Explication II)*:

(Explication II – Event Reduction)* ϕ*-ing* reduces to ψ *-ing* iff

(i) for every *x*, if *x* ϕ*-s* then (x ϕ-s because *x* ψ *-s*), and
(ii) ϕ*-ing* = ψ *-ing*.

Thus, it is not at all clear what the grammatical difference amounts to. To be sure, there are aesthetic and, maybe, pragmatic differences between these question-answer pairs. Maybe, 'because' triggers expectations as to causal explanations, at least if the explanandum deals with an event; and causal explanations cannot be rephrased in terms of 'by'. 'Because' allows for such constructions, 'by' does not (cf. Schnieder 2009). This may suffice to grant a semantic difference between the two. However, a reductive 'by'-explanation can be characterized in terms of a 'because'-explanation plus identity, thereby canceling the possibility that a causal relation backs up the explanation. It seems far-fetched to claim that the question-answer pairs, together with the additional information, differ fundamentally in informational content. At least for practical purposes, knowing

[22] Mulligan (2006) describes a similar explanation as reductive.

that the pumping of blood of the heart occurred because its expanding and contracting regularly occurred, and that the former is identical to the latter, is equivalent to knowing that the heart pumps blood by its component muscles contracting regularly, and the former is identical to the latter. Thus, the difference between 'because' and 'by'-explanations does not suffice to show that reductive 'by'-explanations are informationally or otherwise relevantly different from the corresponding 'because'-explanations plus the relevant identity statement (which excludes the causal interpretation of the 'because' statement).

Note that in the debate on mechanistic explanation, we seem to be confronted with somewhat deviant cases. The basic idea is that, intuitively, two *different* objects may play a role in the explanation – the one being the object that does something which is to be explained, and the other being the object (or objects) which (together) give rise to the target that is to be explained. Here is an example: 'The heart pumps blood by organized contraction of heart muscles.' Can such an explanation be given a straightforward reductive interpretation? The identity clause in a characterization of reductive explanation would commit us to plural identities. Schematically:

1. *F-s φ by the ψ-ing* of *G*-s
2. *a* φ-s by the ψ-ing of $b_1, b_2, b_3, \ldots$

Corresponding reductive interpretations suggests themselves:

(1*) The *φ-ing* of *F-s* reduces to the *ψ-ing(s)* of *G*-s *iff*

(i) *F-s φ by the ψ-ing(s)* of *G*-s &
(ii) The *φ-ing* of *F-s* = the *ψ-ing(s)* of *G*-s.

(2*) *a*'s *φ-ing* reduces to the *ψ-ing* of $b_1, b_2, b_3, \ldots$ *iff*

(i) *a* φ-s by the *ψ-ing(s)* of $b_1, b_2, b_3, \ldots,$ &
(ii) *a*'s *φ-ing* = the *ψ-ing(s)* of $b_1, b_2, b_3, \ldots$

To fully account for this idea, we will need an elaborate notion of plural and partial reduction. A reductionist may wish to argue that the pumping of blood of a heart is *nothing but* the organized contraction of heart muscles (in an appropriate environment). If events are bearer-individuated, then the *identity* clause will require some sort of plural interpretation. This is a substantive issue, and one may want to hold that if this is so, then we should better reject the reductive interpretation of these explanations. A whole is more than its parts, and events a whole engages in are distinct from the events of the whole's constituents. The reductionist commitment is that we can identify the latter with the former. So, an explication of reduction should make this commitment transparent; fortunately, an explication does not have to solve the issue of whether or not such reductions are feasible. Section 5.6.2 proposes a way to accommodate these doubts within a reductionist framework, defining a notion of partial as well as of plural *reduction* without requiring a notion of plural *identity*.

That said, let us turn back to the question of how this form of 'by'-explanation relates to 'because'-explanations. Again, it implies a corresponding 'because'-statement. Schematically, if *F*-s φ (*a* φ-s) by the ψ*-ing* of *G*-s (of b_1, b_2, b_3, ...), *then F*-s φ (*a* φ-s) *because G*-s (b_1, b_2, b_3, ...) ψ. Grasping the 'by'-explanation, we grasp that no causal relation is involved. This is not so for the 'because'-version. However, together with the relevant identity claim, we can define the reductive interpretation of this form of 'by'-explanation in terms of an identity claim and the corresponding 'because'-statement.

Now, 'why' and 'how' questions do not exhaust the field of questions that seek for a reductive explanation. Remember the connection between 'what'-questions and 'by'-explanations tentatively suggested above. Let us push it a bit more: Why shouldn't we say that answering the relevant 'what' questions is, or can be, part of an explanatory project? We may ask what the nature of an entity is and, thereby, seek an explanation. One way of phrasing a 'why'-question is particularly interesting in this respect: Consider the question: 'Why is this water?' It may be interpreted as an *epistemic* question, seeking for a reason to believe that something is water. It can also be uttered to ask for a conceptual clarification – if one needs an update on the conceptual content of 'water'. It can, however, also be asked to arrive at a reductive explanation. Depending on context, an appropriate answer to this question may be this: 'This is water because it is H_2O'. Introducing a contrast, this becomes apparent: 'Why is this water, rather than the watery stuff from twin-earth?' Answer: 'This is water (and it is not the watery stuff from twin earth) *because* it is H_2O, rather than XYZ.' Similar questions could be asked for events. They can be captured as follows:

'Why is this an F/a φ-ing'
'This is an F/a φ*–ing* because it is a G/a ψ-ing.'

A 'what'-question can be used in a similar fashion: 'What is water (ultimately)?' Answer: 'Water is H_2O'. We may also ask what the cause of something was, thereby seeking an explanation. And we may ask what the concept of knowledge is, or what it consists in, thereby seeking a conceptual explanation. On a technical understanding of 'explanation', we do *not* get an explanation when answering such questions: when answering such questions, we do not get sentences that are or express explanations. There just is no term such as 'by', 'in virtue of', 'because' etc. occurring in these sentences. However, it should be obvious that in explanatory projects, 'what' questions may, if appropriately phrased, play a role quite similar to 'how'- and 'why'-questions. The moral to be drawn is this: We should not rely on the grammatical nature of questions, or the occurrence of expressions such as 'by' rather than 'because' in order to determine a kind of explanatory dependence, such as mechanistic or reductive dependence.[23]

Here is an additional argument, namely, that the relevant 'what', 'why', and 'how' questions are *interdependent* in the following sense: they have similar answers whose truth-values stand and fall together, depending on the ontological

[23]For additional examples, see Schnieder (2009), who points out that also evaluative questions may come in the 'how'-format.

make-up of our world. Consider these questions (other versions of ‘how’- and ‘why’-questions would work as well):

1. How does social cognition occur?
2. What is social cognition?
3. Why is this social cognition?

Once we vary the ontological make-up of the world in which these questions arise, the corresponding (correct) answers will take a different form simultaneously:

Let us turn the naturalist reductionist position upside down and let us check the interdependence of truth-values to answers to this set of questions. Assume that we live in a the caricature of a Cartesian world in which our soul, a non-physical substance, directly models other souls when coming into their trans-gravitational field (or whatever does the relevant job); let this event be an event of *soul-modeling*. In this case, we should give a different answer to the first question; How does social cognition occur? By soul-modeling. And what is social cognition? Soul-modeling! The same holds for the third answer: My seeing you as a social being would be an instance of social cognition because it is an instance of soul-modeling. Therefore, varying the relevant bit of ontology of our world, we vary the conditions on answers to these questions. At least to this extent, these questions are connected.

This makes room for broadening the focus. ‘How’-questions are concerned with events. There is no way of phrasing a question about the nature of, say, water in terms of a ‘how’-question (unless the nature of water is fully captured by the events in which water plays a role). Reductive explanation, however, should not only capture ‘by’-explanations, but also explanations such as ‘This is water *because* this is H_2O’, and ‘This is money because we accept it as being money’. Given the tight connection between ‘by’-explanations and ‘because’-explanations, and the idea that even ‘what’-questions can be explanation-seeking, we should not focus primarily on grammatical surface structure in order to individuate the right sort of explanation.

Mechanistic and reductive explanations at least overlap. The fact that mechanistic explanation has been explicated in terms of ‘by’ explanations does not ground a substantive difference between reductive and mechanistic explanation. Ignoring the difference between properties and event-types, the explication proposed in the beginning captures at least some mechanistic explanations (though plural versions still pose a problem):

(Explication I – Kind Reduction) *F-ness* reduces to *G-ness* iff

(i) for every x, if x is F then (x is F *because* x is G), and
(ii) $F\text{-}ness = G\text{-}ness$.

The connection between reductive and mechanistic explanation suggests an interpretation of the directionality of reduction: Intuitively, a mechanistic explanation gives access to the nature of the entity we are dealing with. So does reductive explanation, by presenting one and the same object in different ways.

4.6 Conclusion

We have explicated a core notion of reduction, according to which diversity is conceptual in nature, directionality is explanatory, and unity is cashed out in terms of identity. We are, thus, here:

Q1: How can we reconcile *diversity* and *directionality* with strong *unity*?
...
Th. 3: *Directionality* is explanatory in nature; reductive explanation is a cognate of mechanistic explanation.
Th. 4: In this spirit, a core notion .... can be explicated.
...

The intuitive notion of reductive explanation has been contrasted with a conception of mechanistic explanation. In order to arrive at a cognate of the notion of reduction, we have to abstract away from three features of the notion of mechanistic explanation: we neglect teleology, we do not only focus on 'how'-questions, and we do not only focus on events as the targets of the *explanandum*. The concept of kind-reduction has been characterized as follows (the explanation in (i) is non-conceptual in nature):

(Explication I – Kind Reduction) *F-ness* reduces to *G-ness* iff

(i) for every x, if x is F then (x is F *because* x is G), and
(ii) $F\text{-}ness = G\text{-}ness$.

In the light of the similarity between mechanistic and reductive explanation, we may now tentatively characterize the connection between the *directionality* of reductive explanation and the conceptual *diversity* involved in reductions as follows: *The conceptual difference between the conceptual contents of 'F' and 'G' is a difference in the transparency to the object's constitutive structure, or its nature.*

Thus, reductive explanation differs from characterizations of mechanistic explanation in that it (i) extends the set of targets, (ii) replaces the notion of a mechanism by the broader (though still intuitive) notion of a constitutive structure, (iii) puts emphasis on the fact that there is a whole family of questions tied to this sort of explanation, and (iv) guarantees the reductive interpretation by stressing the identity-requirement. Dropping this identity requirement, we may arrive at a kind of explanation that covers reductive explanation as a special case. Thus, reduction reconciles descriptive or conceptual diversity and explanatory directionality with strong unity, which is cashed out in terms of identity. But can't we get rid of explanatory notions in our characterization of the dependence relation? In the next section, I will suggest that even though we cannot, we can say a lot more about relevant features of the descriptive diversity in virtue of which the explanatory link is instantiated in cases of reduction.

Bibliography

Anscombe, Elizabeth. 1957. *Intention*. Oxford: Basil Blackwell.
Bechtel, William. 1994. Levels of descriptions and explanation in cognitive science. *Minds and Machines* 4: 1–25.
Bechtel, William. 2007. Reducing psychology while maintaining its autonomy via mechanistic explanation. In *The matter of the mind: Philosophical essays on psychology, neuroscience and reduction*, ed. Maurice Schouten and Huib Looren de Jong, 172–198. Oxford: Basil Blackwell.
Bechtel, William. 2008. *Mental mechanisms: Philosophical perspectives on cognitive neuroscience*. London: Routledge.
Bechtel, William, and Adele Abrahamsen. 2005. Explanation: A mechanistic alternative. *Studies in History and Philosophy of the Biological and Biomedical Sciences* 36: 421–441.
Bechtel, William, and Robert Richardson. 1993. Emergent phenomena and complex systems. In *Emergence or reduction? Essays on the prospects of nonreductive physicalism*, ed. Ansgar Beckermann, Hans Flohr, and Jaegwon Kim, 257–288. Berlin: de Gruyter.
Block, Ned. forthcoming. Functional reduction. In *Supervenience in mind: A festschrift for Jaegwon Kim*, ed. Terry Horgan, David Sosa, and Marcelo Sabates. Cambridge, MA: MIT Press.
Byrnes, W.C., and C.A. Jensen. 2001. Exercise physiology. In *Foundations of exercise science*, ed. Gary Kamen, 15–32. Lippincott: Williams & Wilkins.
Causey, Robert. 1977. *Unity of science*. Dordrecht: Reidel.
Chalmers, David. 1996. *The conscious mind*. Oxford: Oxford University Press.
Crane, Tim. 2001. *Elements of mind*. Oxford: Oxford University Press.
Craver, Carl. 2001. Role functions, mechanisms and hierarchy. *Philosophy of Science* 68: 31–55.
Craver, Carl. 2005. Beyond reduction: Mechanisms, multifield integration and the unity of neuroscience. *Studies in the History and Philosophy of Biological and Biomedical Sciences* 36: 373–395.
Craver, Carl. 2006. When mechanistic models explain. *Synthese* 153: 355–376.
Craver, Carl. 2007. *Explaining the brain: Mechanisms and the mosaic unity of neuroscience*. Oxford: Oxford University Press/Clarendon Press.
Cummins, Robert. 1975. Functional analysis. *Journal of Philosophy* 72: 741–765.
Cummins, Robert. 1983. *The nature of psychological explanation*. Cambridge, MA: Bradford/MIT Press.
Cummins, Robert. 2000. How does it work? versus 'what are the laws?': Two conceptions of psychological. In *Explanation and cognition*, ed. F. Keil and R. Wilson, 117–144. Cambridge, MA: MIT Press.
Darden, Lindley, and Nancy Maull. 1977. Interfield theories. *Philosophy of Science* 44: 43–64.
Davidson, Donald. 1963. Actions, reasons and causes. *Journal of Philosophy* 60: 685–700.
Davidson, Donald. 1967. The logical form of action sentences. In *The logic of decision and action*, ed. Nicholas Rescher, 81–120. Pittsburgh: University of Pittsburgh Press.
Davidson, Donald. 1969. The individuation of events. In *Essays in honor of Carl G. Hempel*, ed. Nicholas Rescher, 216–234. Dordrecht: Reidel.
Dennett, Daniel. 1994. Cognitive science as reverse engineering: Several meanings of 'top down' and 'bottom up'. In *Logic, methodology and philosophy of science IX*, ed. D. Prawitz, B. Skyrms, and D. Westerståhl, 679–689. Oxford: Elsevier Science.
Doerschuk, Peter C., Robert R. Tenney, and Alan S. Willsky. 1977. Modelling electrocardiograms using interacting Marcov chains. *International Journal of Systems Science* 21: 257–283.
Feinberg, Joel. 1970. *Doing and deserving. Essays in the theory of responsibility*. Princeton: Princeton University Press.

Fodor, Jerry A. 1974. Special sciences: Or the disunity of science as a working hypothesis. *Synthese* 28: 97–115.
Friedman, Kenneth. 1982. Is intertheoretic reduction feasible? *The British Journal for the Philosophy of Science* 33: 17–40.
Glennan, Stuart. 1996. Mechanisms and the nature of causation. *Erkenntnis* 44: 49–71.
Glennan, Stuart. 2002. Contextual unanimity and the units of selection problem. *Philosophy of Science* 69: 118–137.
Goldman, Alvin. 1970. *A theory of human action*. Princeton: Princeton University Press.
Kauffman, Stuart A. 1970. Articulation of parts explanation in biology and the rational search for them. In *PSA: Proceedings of the biennial meeting of the Philosophy of Science Association*, 257–272. Dordrecht: Reidel.
Kim, Jaegwon. 1976. Events as property exemplifications. In *Action theory*, ed. Douglas Walton and Myles Brand, 159–177. Dordrecht: Reidel.
Kim, Jaegwon. 2002. The layered model: Metaphysical considerations. *Philosophical Explorations* 5: 2–20.
Kim, Jaegwon. 2008. Reduction and reductive explanation. Is one possible without the other? In *Being reduced. New essays on reduction, explanation and causation*, ed. Jesper Kallestrup and Jakob Hohwy, 93–114. Oxford: Oxford University Press.
Kitcher, Philip. 1981. Explanatory unification. *Philosophy of Science* 48: 507–531.
Kitcher, Philip. 1989. Explanatory unification and the causal structure of the world. In *Scientific explanation*, ed. Wesley Salmon and Philip Kitcher, 410–505. Minneapolis: University of Minnesota Press.
Kuhn, Thomas. 1962. The Structure of Scientific Revolutions. Chicago: University of Chicago Press.
Lakatos, Imre. 1977. *The methodology of scientific research programmes: Philosophical papers*, vol. 1. Cambridge: Cambridge University Press.
Machamer, Peter, Lindley Darden, and Carl Craver. 2000. Thinking about mechanisms. *Philosophy of Science* 67: 1–25.
McMullin, Ernan. 1978. Structural explanation. *American Philosophical Quarterly* 15: 139–147.
Melnyk, Andrew. 2003. *A physicalist manifesto: Thoroughly modern materialism*. Cambridge: Cambridge University Press.
Mulligan, Kevin. 2006. Ascent, propositions and other formal objects. *Grazer Philosophische Studien* 72: 29–48.
Nagel, Ernest. 1935. The logic of reduction in the sciences. *Erkenntnis* 5: 46–52.
Nagel, Ernest. 1961. *The structure of science. Problems in the logic of explanation*. New York: Harcourt, Brace & World, Inc.
Oppenheim, Paul, and Hilary Putnam. 1958. The unity of science as a working hypothesis. In *Concepts, theories, and the mind-body problem*, ed. Grover Maxwell, Herbert Feigl, and Michael Scriven, 3–36. Minneapolis: Minnesota University Press.
Pearl, Judea. 2000. *Causality: Models, reasoning, and inference*. Cambridge: Cambridge University Press.
Pettit, Philip, and Frank Jackson. 1992. Structural explanation in social theory. In *Reduction, explanation, and realism*, ed. Kathleen Lennon and David Charles, 97–132. Oxford: Oxford University Press.
Salmon, Wesley. 1984. *Scientific explanation and the causal structure of the world*. Princeton: Princeton University Press.
Schaffner, Kenneth. 1993. *Discovery and explanation in biology and medicine*. Chicago: The University of Chicago Press.
Schaffner, Kenneth. 2006. Reduction: The Cheshire Cat problem and a return to roots. *Synthese* 151: 377–402.
Schnieder, Benjamin. 2009. A refutation of the anscombe thesis. *Linguistics and Philosophy* 31: 649–669.
Searle, John. 2010. *Making the social world. The structure of human civilization*. Oxford: Oxford University Press.

Simon, Herbert. 1969. *The sciences of the artificial*. Cambridge, MA: MIT Press.
Sneed, Joseph. 1994. Structural explanation. In *Philosophy of physics, theory structure and measurement*, ed. Paul Humphreys, 195–213. Dordrecht: Kluwer.
van Riel, Raphael. 2011. Nagelian reduction beyond the Nagel-model. *Philosophy of Science* 78: 353–375.
Wimsatt, William C. 1976. Reductive explanation: A functional account. In *PSA: Proceedings of the biennial meeting of the Philosophy of Science Association*, 1974, ed. G. Pearce, A.C. Michalos, C.A. Hooker, and R.S. Cohen, 671–710. Dordrecht: Reidel.
Wimsatt, William C. 2006. Aggregate, composed, and evolved systems: Reductionistic heuristics as means to more holistic theories. *Biology and Philosophy* 21: 667–702.
Wimsatt, William C. 1997. Aggregativity: Reductive heuristics for finding emergence. In *Philosophy of science, supplement: Proceedings of the 1996 biennial meetings of the Philosophy of Science Association, Part II: Symposia papers* 64: S372–S384.
Woodward, James. 2003. *Making things happen. A theory of causal explanation*. Oxford: Oxford University Press.

Chapter 5
Reductive Explanation

Reduction reconciles diversity and directionality with strong unity. Unity is cashed out in terms of identity. Diversity is descriptive, or conceptual. Directionality is the directionality of explanatory dependence. A sentence of the form '*F-ness* reduces to *G-ness*' expresses a truth if and only if (i) for every x, if x is F then (x is F *because* x is G), and (ii) F*-ness* = G*-ness*; the explanation in condition (i) is non-conceptual. The direction of the dependence relation depends upon features of the descriptions under which an object is presented by the expressions flanking the reduction predicate. It goes from *water* to H_2O, but not from H_2O to *water*. This is so because of a difference in semantic facts about 'water' and 'H_2O', facts other than those regarding reference or designation. Recall: Expressions do not matter in this respect. Similarly, ontology does not matter (water just is H_2O).

Is this all there is to say about reduction? It is not. In this chapter, I will give a more thorough idea of how descriptive or conceptual diversity is tied to the directionality of reduction.

I will suggest that the explanatory directionality of reduction is tied to differences in how different descriptive or conceptual contents, or Fregean senses present us with entities *as having different properties*. The idea is this: Re-describing an object, we may get better access to its nature. Directionality of reduction just is directionality that stems from a difference in the degree to which conceptual or descriptive contents give access to an objects nature, or its constitutive structure. This is reductive dependence. Intuitively, water reduces to H_2O because 'H_2O' presents us with water under properties that differ relevantly from those under which 'water' presents us with water. The present section proposes an account of this difference in the sense that it offers a semantic model that enables us to tie reductive dependence to specific semantic features of expressions that pick out a reduced or a reducing object. The difference just is the difference tracked by the directionality of explanatory dependence. Thus, I will not come up with a "reductive" account of reductive explanation; explanatory notions will occur in the characterizations of the central notions of the theory proposed here.

In a first step, I will propose an intuitive outline of the model (Sect. 5.1). I will then introduce the details (Sects. 5.2, 5.3 and 5.4) and argue that a mereological

R. van Riel, *The Concept of Reduction*, Philosophical Studies Series 121,
DOI 10.1007/978-3-319-04162-9_5, © Springer International Publishing Switzerland 2014

interpretation of reductive dependence is bound to fail (Sect. 5.5). I will then introduce a more thorough version of *Explication I*, show that the explication meets the job description, and characterize derivative notions (Sect. 5.6). Finally, it will be shown how the proposal offered here provides a structure for explications that exploit other resources (Sect. 5.7).

5.1 The Model: An Intuitive Outline

Things have, on the view proposed here, properties *according to* a description. The properties an entity has according to a description are captured by the technical notion of a *property structure*. Some expressions give access to *constitutive property structures*, a special case of property structures, of the entities they designate. A constitutive property structure gives, to some extent, access to the object's nature. Different constitutive property structures differ in the degree to which they give access to an object's nature. Consider, again, 'water' and 'H_2O' (here treated as an *empirical formula*, lacking information about structural arrangement, to avoid complexity). The designatum of 'H_2O' is, when presented under the concept of H_2O, presented as *the entity instances of which exemplify the property of having exactly three atomic constituents, two of which exemplify the property of being hydrogen, and one of which exemplifies the property of being oxygen.* In this, the meaning of 'H_2O' relevantly differs from the meaning of 'water'. If 'water' has descriptive content, describable, say, in terms of quenching thirst and being tasteless, then the meaning of 'water' presents us with instances of water as quenching thirst and being tasteless. If 'water' lacks descriptive content, then instances of water are presented as being water by 'water' – that's it. Thus, the concepts of H_2O and water present us with water under different property structures. A property structure is an abstract object an expression E gives access to, which contains exactly those properties the entity designated or signified by E has *according* to E's conceptual content, or meaning. The directionality of reductive explanation tracks relevant differences in property structures that present one and the same object.

In order to cash out this idea, we need a precise idea of the following notions: that of (i) a property structure, (ii) a constitutive property structure, and (iii) of having properties according to a description.

Here is the short, intuitive version: a property structure is a tuple of properties and functions defined over properties. A constitutive property structure is a property structure that contains constitutive properties of the entity so described or presented; the notion of *being a constitutive property of* is here understood as an explanatory notion. And an object *x has specific properties according to the meaning or conceptual content of an expression E* iff according to the proposition that E adequately describes an object, y, y has these properties, and $x = y$. Before giving a more thorough interpretation of these and related notions, let me address two questions that may seem to suggest themselves: (i) what is the supposed relation between property-structures and meanings or conceptual contents? And (ii) shouldn't we try to get rid of expressions such as 'in virtue of' and 'because' in a characterization of reduction?

5.1.1 Property Structures and Semantics

It is important to note that talk about property structures is supposed to model *one aspect* of (non-Fregean) meaning, (Fregean) sense, or conceptual content instead of fully capturing this idea. There might be more to the meaning, sense, or conceptual content of the *explanans* and the *explanandum* of reductive explanation, or to the meaning of instances of '*a*' and '*b*' in a sentence of the form '*a* reduces to *b*' than what is explicit in property structure talk. There are two ways of interpreting this distinction. On the one hand, we might adopt *semantic pluralism*, according to which different semantic values of an expression may play a role in different contexts. The relevant semantic values may shift from context to context.[1] This idea obviously goes back to Frege, who himself introduced different sorts of semantic values, or semantic relations. Thus, one might want to be liberal and hold that property structures are one semantic value of expressions, one amongst others – the one that kicks in, for example, when we want to understand reductive statements.

Alternatively, we could just say that whatever the semantic value of an expression actually is, it *determines* a property structure; in this sense, we are able to model an expression's semantic value in terms of a property-structure, and it is this aspect which helps illuminating the semantic condition on instances of reductive relations. I will adopt this latter, less committal interpretation. This interpretation suits the purpose of arriving at an explication: An explication should be sparse in the sense that it should not invoke assumptions that are not required to explicate the target concept. Later, I will suggest that the explication could be given a two-dimensionalist interpretation (see Sect. 5.7).[2] Whatever meanings are, the reductionist is committed to the idea that meanings determine or correspond to something like property structures, and play a role similar to the role these structures play within the explication offered below.

It is sometimes assumed that constituents of meanings are concepts rather than properties or individuals. Schiffer characterizes the relation between Fregean propositions (as meanings of sentences) and Russellian conceptions of propositions,

[1]David Chalmers (2004) suggests that epistemologically explicated notions of meaning may play a role in some contexts, whereas metaphysically explicated notions may play a role in others, and none of these is more fundamental than the other. See Sect. 5.7 for a discussion. Similarly, Albert Newen introduces meanings as vector spaces (1996), which comprise several semantic values for expressions.

[2]Apparently, it could also be phrased in hyper-intensional semantics such as those proposed by Tichy (1988) and Cresswell (1985), and also Church (1973, 1974, 1993).

writing that "[j]ust as the Russellian may take $<<x_1, \ldots, x_n>R^n>$ to represent the form of any proposition, so the Fregean may take it to be represented by $<<c_1, \ldots, c_n>C^n>$" (Schiffer 2003, 22), where the building blocks of propositions are *concepts* of properties or objects, rather than properties or objects themselves.[3] The interpretation given here does not contradict the assumption that there might be meanings the building blocks of which are *not* properties (or individuals), but rather concepts of these (or, in Fregean terms: *senses*). But even on the Fregean interpretation, we should assume that meanings or conceptual contents give access to property structures. From a sentence of the form '*Fa*' we can infer that *a* exemplifies the property of being *F*. According to Strawson (1974, 33) and Quine (1980, 164) these are synonymous.[4] Predicates are used to attribute properties, at least sometimes, and, under a semantically realist interpretation of science, at least in science. Thus, it is quite natural to accept that simple predicates and kind terms give access to properties, and that semantically complex expressions give access to things as having properties. But what is the difference? Maybe there is none. This is a topic that is largely irrelevant for the present purpose. The important point is this: Properties are the things in virtue of which entities that exemplify these properties behave the way they do. Properties are not representational in nature. If concepts are representational in nature, then there is an important difference. But, one might want to ask, what are properties?

Let me respond with this dialectical remark instead of a straightforward answer. In Sect. 2.4 I mentioned several possibilities to conceive of properties. The account proposed here is compatible with all of these interpretations. So, I rely on an intuitive notion of a property, a notion that is partly characterized as follows: Properties do not belong to the representational furniture of our world, and they form the class of things some of which are such that in virtue of them, space-time objects behave the way they do. We might go a step further and claim, with Bealer (1993, 20), that we should take properties at *face value*, that is: the category of properties should be treated as a category which is not to be rephrased in, say, set-theoretical terminology. From a dialectical point of view, we could adopt this assumption: Within the account proposed here, the notion of a property is interpreted as basic. But what about explanatory concepts?

---

[3] If we assume that property-structures are semantic values, we might want to hold that the account given here is a version of Russellian semantics. At first sight, one might hope that, building on an understanding of a sense as a mode of presentation, one could describe Fregean senses in terms of property-structures. For a critical examination of the notion of a mode of presentation, see (Künne 2001).

[4] Schnieder (2006) argues that this thesis is false, because there is a difference in understanding conditions between sentences explicitly attributing properties and those that rely on the 'predication-mode'. For present purposes, this difference does not matter; all that is required is the assumption that the inference is correct.

5.1.2 *Treating 'Because' as Basic*

The notion of a constitutive property structure will be cashed out in explanatory terminology. Obviously, one might feel tempted to try to eliminate expressions such as 'in virtue of', 'because', and 'by' from the model of reduction. However, as the long-lasting discussion on related topics, such as supervenience and ontological dependence shows (for an overview, see Correia 2008), these attempts are not very promising. Modal notions do not help to perform the relevant trick of defining a relevantly directional relation (unless we *stipulate* directionality). I will not try to get rid of such expressions. Following a recent trend in related domains, I suggest that we take directionality at *face value*. However, saying that explanatory notions cannot be analyzed away is *not* to say that there isn't anything interesting to say about them; in our case, it is not to say that there isn't anything interesting to say about the connection between descriptive plurality and explanatory directionality. In this sense, I will introduce a model that enables us to give a precise idea of the semantic conditions the *explanans* and the *explanandum* of a reductive explanation have to meet in order to enter a true instance of this type of explanation.[5]

Ideally, to obtain an expression that gives us the relevant property structure for another expression, we have to rely on a meaning revealing definition of the latter. From the definiens of such a definition, we can obtain a (usually complex) expression that contains constituents that designate the relevant building blocks of property-structures. Intuitively, this feature makes them transparent with respect to the property-structure's organization. In the sections to follow, we can then apply an operator to the resulting expression, which designates the property-structure itself, rather than the designatum or significatum of the expression we started with. This can be conceived of as a procedure consisting of two individual steps to obtain a term that is transparent with respect to the building blocks of the property structure it designates – a procedure to obtain a property structure term. Having introduced this procedure, we can give an idea of what it is for a thing to be a certain way according to a description. We will reflect upon the metaphysics of property structures and define a number of notions relevant to account for reduction. Building on these definitions, we can then suggest a more thorough explication of a core notion of

---

[5]This is not the common way to treat issues of explanation, at least in the philosophy of science. Often, kinds of explanations are described with respect to how they relate to pragmatic issues (van Fraassen 1980), to epistemological issues (according to the interpretation of Covering-Law models of explanation as being tied to the form of arguments), and to epistemic procedures of discovery (Craver 2007; Bechtel 2007) and intervention (Woodward 2003; Craver 2005). Types of explanations can be individuated in terms of characterizability within a specified framework (we then arrive at characterizations such as 'explanations that fit the DN-model') or in terms of how they relate to causation or other ontological relations (we then arrive at characterizations such as 'causal explanation'). However, it should be clear that this is a way to account for explanation; learning something about the conditions on the semantics of explanans and explanandum (conditions fixing when these are appropriate), we learn a lot about this sort of explanation.

reduction, defend this explication's adequacy, define derivative notions and point to alternative conceptions of those objects that play the role property structures play in the present account.

5.2 Meaning-Revealing Definitions and Property-Structures

Basically, a property structure (henceforth, 'PS') consists of properties and, sometimes, functions that map properties or states of affairs onto properties or states of affairs. PSs will be conceived of as being fully determined by the PSs an expression's semantically simple constituents give access to and these constituents' arrangement (this mimics compositionality). This will base a procedure to *obtain* expressions that refer to PSs in an interesting way – in a way that makes the arrangement of properties and functions in a PS transparent. Thus, let us, in a first step, reflect upon which expressions come in an appropriate format to enable us to grasp the property structures they give access to. In the case of complex expressions that consist of semantically simple expressions *only* (and for semantically simple expressions themselves) we can easily arrive at the basic constituents of the PS this expression gives access to. Moreover, the syntax of such an expression will give access to the arrangement of the properties within this structure. For semantically complex expressions that are syntactically simple, we have to rely on characterizations that illuminate these expressions' meanings. Assuming that ultimately, these characterizations will contain semantically simple expressions only, we can, again, easily identify the property-structure's constituents and their arrangement. Obviously, relying on expressions that contain semantically simple expressions *only* is an idealization. However, it will help to make the idea precise.

Consider again 'H_2O' and assume that each of its meaningful constituents is semantically simple. Intuitively, it gives access to a PS that contains, among other properties, the property of being a proper part of (recall that relations are here treated as n-ary properties, with $n > 1$), the property of being hydrogen and the property of being oxygen. This PS can easily be obtained from the predicate '_is H_2O', or one of its ordinary language reformulations. The basic constituents of the PS H_2O gives access to are the properties signified by predicates or designated by terms which, together, form the complex term 'H_2O' (or one of its ordinary language reformulations), and rules associated with the formation of this term.

What about syntactically simple, semantically complex expressions? Assume that 'water' means 'being tasteless and being liquid'. In this case, the constituents of the property structure 'water' gives access to are not obtainable from constituents of 'water'; rather, they are obtainable from an *appropriate analysis or characterization of the concept of water*.[6] Thus, sometimes, to account for the PS an expression gives

[6]For the proposed account to work, we do not have to rely on a too demanding sense of 'analysis' or 'meaning-revealing definition', i.e. on a sense that suggests that an appropriate definiens and its

access to, we have to rely on characterizations of the meaning of that expression; this will be the case iff the expression does not fully reveal the relevant semantic structure.

Let me comment upon several sorts of meaning-characterizations, conceived of as definitions, pertinent in the context of reduction to make this idea more precise. We will then be able to build upon the definiens of such a definition (or the counterpart in a meaning-characterization of a non-predicate, like a term or a sentence) of an expression E to specify the PS under which E presents its designatum (if it is a term) or significatum (if it is a predicate or a sentence). There are at least three ways we can treat the complex meaning of an expressions, which might be relevant in the case of reduction: the meaning of functional terms, of directly referential terms, and of alleged cases of non-functional terms. The next Sect. (5.2.1) comments on these cases, before we turn to the idea that meaning revealing definitions can differ in the degree of explicitness (Sect. 5.2.2). Building on the notion of a maximally explicit definition, we are in a position to give an idea of how we get access to property-structures.

5.2.1 Arriving at the Meaning of Expressions

If functional terms gain their meaning by the theory they are used in, the property structure they reveal will (at least partly) be given via a definition building on the Ramsey sentence of the theory the term occurs in (if you are not familiar with the notion, please read the first few paragraphs in Sect. 6.7). Thus, to appropriately model the property structures of the terms in 'pain reduces to C-fiber stimulation', we should not model the property structure revealed by 'pain' as *being pain*. Assume that pain is exhaustively described by being the internal state that maps tissue

definiendum must be strictly synonymous, although this is the way the account is presented here. To bypass problems concerning the analytic-synthetic distinction (Quine 1951) or the paradox of analysis (Black 1944), we could rely on a notion weaker than that of synonymy or analyticity; just substitute it by the concept of a functionally appropriate explication. Building on some such idea, we could rely on a notion of *meaning similarity*, or *similarity in use*. But don't we have to give a precise idea of how this is supposed to work? We do not. The job-description of the technical term reduction does, I suggest, commit us to the idea that meanings present us with an object in a specific way. An appropriate model of reduction is bound to give an idea of how meanings manage to do this. Thereby, any appropriate model of reduction is committed to the assumption that different meanings present us with objects in different ways. One way of cashing out a difference in meaning is this: Two expressions have different meanings (in the relevant sense) iff they have different illuminating definitions. An appropriate definiendum gives access to the relevant property structure under which the definiens presents us with an object. But what is an appropriate definiendum? One way of cashing out this idea is in terms of analyticity or synonymy. An alternative explication might build upon the idea that the appropriateness of a definiendum, even for non-technical expressions, is to be cashed out in terms of explication, or something similar, rather than analyticity or synonymy. The overall-strategy, however, will remain the same.

damage onto yelling (thus, assume a very simplified version of folk-psychology). Assume, moreover, that the concepts of *tissue damage* and *yelling* are not further analyzable. Then, the functional concept of pain is fully given by *being such that any instance of it is caused by tissue damage and causes a yelling*. From this, we can obtain the relevant property structure. Thus, for functional terms, there is a well-established procedure to obtain the relevant definitions. But what if functional definitions are not always available?

Directly referential expressions lack descriptive content, as Kripke famously argued (Kripke 1980). A number of expressions that play a prominent role in reductions may turn out to be directly referential, such as 'water', 'iron', and numerous other expressions we use in extra-scientific discourse to talk about our environment. That is: these expressions may have a referent or signify something, but, at the same time, lack meaning (in the sense the term 'meaning' is employed here). I propose the following treatment of these cases: If, say, 'water' lacks meaning but designates the property of being water, then it presents us with water *as being water*. Note that this treatment is insensitive to a possible difference between semantically simple expressions that designate properties under a simple concept and directly referential expressions. This will not matter much for the discussion to follow.

Assume that some ordinary-language expressions, like qualia-terms or, maybe, expressions like 'knowledge', have a meaning that is not functional in nature and can be revealed only via conceptual analysis that does not explicitly rely on a Ramsey-sentence, or, in Chalmer's terms, by reflection on meaning, or ideal rational reflection only (Chalmers 2002, 2004, 2006). Maybe, for some notions, this is the way to go. Note, however, that possibly problematic cases of conceptual analysis will, under some widely shared assumptions, hardly play any crucial role at all in reduction: If natural kind terms lack meaning or if their meaning can be given by functional characterizations, and if it is either natural kind terms or scientific terms (including functional terms) which flank '_reduces to_' in true reduction statements, then the account proposed here will, at least for the relevant kinds of reduction, just not be concerned with issues of *conceptual analysis* of non-functionally analyzable ordinary language expressions; it will merely be concerned with natural kind terms, functional terms and scientific terms.

5.2.2 An Ideal Demand: Maximal Explicitness of a Definition

Now, meaning revealing definitions are often hard to arrive at. So, we should regard the condition that we have to build upon definitions that fully reveal the meaning of the expression we are interested in as an ideal demand rather than a strict requirement needed to make sense of the idea of a PS. Explicit descriptions of PSs can vary in degree of specificity. Let us return to the pain-example to illustrate this point. Assume that the meaning of 'yelling' is not further analyzable, but that the meaning of 'tissue damage' is, say, in terms of *rearrangement of tissue-structures such that due to this rearrangement, these structures cannot fulfill their primary biological function anymore*. Then, we could give a more precise analysis saying that pain is the property that is such that its instances are caused by events that consist of a rearrangement of tissue-structures such that due to this rearrangement, these structures cannot fulfill their primary biological function anymore, and they cause yelling. Thus, definitions come in *degrees* of illumination – they possibly contain expressions the meanings of which could be further analyzed. We can then *compare any two definitions D^1 and D^2 of a predicate 'F_' with respect to which of the predicates occurring in the definiens are further analyzable and which are not.* Similarly, if the PS 'knowledge' gives access to is tied to a characterization of the meaning of 'knowledge' in terms of *being such that any instance is a belief, is true, and is justified*, then it contains properties presented under concepts that are not semantically basic, although the expressions are syntactically basic. The meaning of 'belief' would, according to a functional interpretation of mental terms, be given by a definition obtained from the relevant Ramsey sentence. Any definition of knowledge, which contains a meaning revealing definition of 'belief' in its definiens, instead of 'belief' itself, would be finer grained than the original definition in terms of belief, truth, and justification; more precisely, it would be finer grained with respect to the meaning of 'belief' (or 'is a belief', or 'being a belief'). So, what one might call the problem of granularity is to be solved by allowing for differences in how the definitions reveal the meanings of the terms defined. For the present purpose, we should assume that in ideal cases, we are able to give a full-blown definition, which does not contain predicates or terms that are further analyzable. This can be captured as follows: *An appropriate characterization of the meaning of a term E is maximally explicit iff it does not contain a constituent that is further analyzable.* However, in practice, we do not always have to go all the way down (if there is such a way) to a full-blown definition, which does not leave open one single conceptual issue. In the present context, it suffices to show that for some description in an *explanans* of a reductive explanation, there is a definition that reveals *one* constitutive property structure it gives access to, such that this constitutive property structure can be judged to be more fundamental than the PS given in the *explanandum*. The role of property structures in a characterization of reduction could thus be fulfilled even if an expression gives access to more than one property structure.

To sum up: a PS presents us with entities (if any) as exemplifying (or being related to objects exemplifying) specific properties. Assume that 'water' cannot be further analyzed. Then, it is associated with a property structure that presents us with water as being water. 'H_2O', however, is associated with a property structure that presents us with water as being constituted by exactly three atoms, two of which instantiate the property of being hydrogen, whereas one instantiates the property of being oxygen. The basis for obtaining property structures is, ideally, a definition that fully reveals the meaning of the relevant expression. In the ideal case, we would rely on analyses or characterizations of meanings that are maximally explicit. The notion of analyzability should here be understood in a wide sense, such that any sort of illuminating definition is captured. A characterization can either be a definition of a predicate or a characterization of the meaning of a term or a sentence, which signifies a state of affairs.

Building on this notion, we can give a characterization of a property structure an expression gives access to. *A property structure an expression E gives access to consists of exactly those properties the syntactically simple predicates and terms of a maximally explicit characterization of the meaning of E signify or designate, and of functions working on these properties that fix the property structure's structure in a way determined by the structure of the maximally explicit characterization of E.*

Note that according to this interpretation, we should allow for some expressions to be the appropriate characterizations of their own meaning. This is the case iff these expressions do not contain constituents that are further analyzable. Now, one might object that there are no maximally explicit definitions. Meaning just does not behave that way. If this were correct, we could easily weaken the notion of a property structure an expression gives access to as follows: Which property structure an expression gives access to might be *contextualized* to *different meaning revealing definitions*. Thus, expressions do not give access to property structures *simpliciter*, but rather with respect to meaning revealing characterizations. For the sake of simplicity, I will stick to the assumption that for every expression, there is one maximally explicit meaning-characterization. But how do we arrive at the relevant properties that constitute a PS, based on a given definiens?

5.3 Property Structure Terms

Here is a simple procedure for obtaining expressions that designate property structures from predicates or general terms. This procedure constitutes the first step to obtain property-structure terms, i.e. terms that are transparent with respect to the property-structure a corresponding expression gives access to. To begin with, consider unary predicates that are syntactically simple in the sense that they do not contain other predicates as constituents. For example, from '_is water' we obtain the term 'the property of being water' that designates the property the predicate signifies. Similarly for kinds of events (which are here treated as properties): from '_ is social cognition', we obtain 'the property of being/the event type of social cognition'. For n-ary properties with n > 1 (that is: relations), we have to follow a slightly different procedure. For example, what the two-place predicate '_recognizes_' signifies can be designated by 'the relation of recognition'. Similarly, for three-place predicates: the relation signified by '_ lies in between _ and _' can be designated by 'the relation of lying in between _ and _'. Sometimes, it is worth mentioning the number of relata; consider the two predicates '_ lies in the intersection of _ and _' and '_lies in the intersection of _ and _ and _'. We would get 'the relation of lying in the intersection of two entities' and 'the relation of lying in the intersection of three entities'. (Note that these expressions are ambiguous. Take the first as an example: According to one sense, it designates the property signified by 'there are two entities, _ lies in the intersection of'; this is a *unary property* rather than a relation.) Now, entities do not possess or exemplify relations; rather, they *stand in* relations to other objects.

Using these and similar operations, we are, at least in principle, able to transform any expression that employs the predication-mode to attribute properties into expressions that explicitly attribute properties only. In order to do so appropriately, these operations should be applied to simple constituents of expressions; it would miss the point to treat the predicate '_has instances which have exactly three constituents, two of which are hydrogen and one of which is oxygen' as a simple predicate, expressing a unary property, although it does. Rather, we should apply the operations to constituents that *lack* (semantically relevant) syntactical parts. Thus, they should be applied to simple syntactic constituents – constituents which themselves do not contain predicates or terms designating properties.

5.3.1 Building Constituents of Property Structure Terms

Since natural language is quite flexible, we have to idealize. In order to avoid complexity, I will ignore quantified phrases, definite descriptions and expressions which contain n-ary relations with $n > 2$. Here are the relevant operations on constituents of an expression *E* which contains expressions from a small fragment of English, namely, simple predicates of an arity $n < 3$, singular terms, and connectives like 'and', 'not' and the like:

1. For general terms *F* or *R* and for simple predicates '_is *F*' or '_*R*_' which do not take arguments in an expression *E* and occur in *E*, introduce 'being an *F*' or 'the *R*-relation'.
2. For any simple one-place predicate which takes an argument in *E*, '*a* is *F*', occurring in *E* introduce '*a* exemplifies the property of being *F*'.
3. For any simple two-place predicate that takes one argument in *E* as follows '*a* *R*_', introduce '*a* stands in relation *R* to _'.
4. For any simple two-place predicate that takes two arguments in *E*, introduce '*a* stands in relation *R* to *b*'.
5. For any simple two place predicate which takes one argument as follows: '_*Rb*' introduce '_stands in relation *R* to *b*'.

This procedure will give us the collection of properties the property structure *E* gives access to consists of: it consists of properties referred to by constituents of the resulting expression. The syntactic structure of *E* will have changed slightly; however, the positions of connectives like 'and', and quantifiers and so forth will not have been altered. Let me comment upon the contribution of such expressions to the property structure a complex expression containing some of these elements gives access to. This will be an intuitive outline, rather than an appropriate definition. In the Appendix, a more precise interpretation of the contribution of 'and' and similar expressions will be offered. For the present purpose, however, an intuitive understanding is sufficient.

5.3.2 *The Role of Connectives and Quantifiers*

The *syntactic* operation of generating a term that is disjunctive but nevertheless takes the form of a singular term ('the property which is such that it is either realized by *F* or is realized by *G* or is realized by ...') is beyond suspicion. However, following syntactic rules is not to follow ontological ones. But sometimes, complex constructions do refer to or signify something: Intuitively, '*Fx* and *Gx*' may signify the *property of being F and G*. For example, 'the property of having exactly one hydrogen atom and two oxygen atoms' *refers*, although it is a complex term. But what does 'and' contribute to the property structure of expressions of the form 'The property of being *F* and *G*' (or more complex expressions)? Clearly, 'being red and being green' differs from 'being red or being green'. Following a tradition in the philosophy of language, I suggest treating the contribution of 'and', 'or' and similar expressions, like quantifiers, in this context as expressing functions that map properties or states of affairs onto properties, or states of affairs. For example, in 'being red and being green', 'and' expresses a function that maps the property of being red and the property of being green onto the property of being red and green (if there is such a property). Similarly, for quantifiers: The loving relation is not to be confused with the property of there being someone one loves. The former is signified by '_loves_', the latter by 'there is someone _ loves'. Here, the existential quantifier is treated as expressing a function that maps the binary property (the loving relation) onto the unary property of there being someone one loves. Similarly, applying 'there is an *x* such that' to '*z* loves *x* more than *y*', we get a new predicate, namely: 'there is an *x* such that *z* loves *x* more than *y*'. Here, the quantifier maps a three-place relation onto a binary one. I suggest following this idea of treating the relevant expressions as determining such functions, with two qualifications to accommodate doubts which are pretty frequent in the less Platonistic areas of the reduction debate: The functions will be regarded as *partial* functions, and we do not take natural language expressions like 'and', 'there is at least one' and so forth to *express* such functions, but merely to determine these functions. Let me explain.

Properties and states of affairs and tuples of these form the domain of these functions. If the result of the application of a quantifier is a sentence, then we should regard it as signifying *the state of affairs we can signify (or a fact we can allegedly state, if it holds) using that sentence.* If the result of the application to an n-place predicate signifying an n-ary property (with $n > 2$) or relation is an $n - 1$ place predicate, it yields the $n - 1$-ary property or relation that is signified by the resulting predicate (if this predicate signifies any property at all). If 'and' is applied to sentences '*p*' and '*q*' then it takes the pair of states of affairs *that p* and *that q* as an argument and yields the state of affairs *that p and q*. If it is applied to a predicate '_is *F*' and a sentence '*q*', it takes the pair *<the property of being an F, that q>* as an argument and yields the property signified by the open sentence so generated (if this open sentence signifies a property at all). So it does when applied to two predicates: It takes the pair of properties signified by these predicates as arguments and yields the property signified by the complex predicate so generated (if this predicate signifies a property at all). So does 'not' with states or properties, and other connectives that are mimicked by sentential connectives in predicate logic.[7]

Thus, these functions sometimes yield a new entity, and sometimes they do not. This depends upon which properties actually exist, and which facts actually obtain. Thus, we interpret them as partial functions and specify that if they assign a value to an argument (we do not know when this is the case) how they do this. This captures the idea that some disjunctive property-expressions may signify or refer to properties, whereas others do not (see Sect. 6.5.2). But how do these functions relate to the natural language expressions 'and', 'or' and natural language quantifiers? If we were to give a formal treatment of these expressions, we could assign these functions as these expressions' semantic values. However, we can easily be a little more cautious here without making the idea useless.

To illustrate the strategy, let us focus on 'and'; similar expressions can be treated in a similar fashion. Let the natural language meaning of 'and' in instances of an expression of the form 'being F and G' be *[and]**. Let the function associated with 'and' described above be f^{and}. According to the strong interpretation, which is the basis for formal language treatments of such natural language issues, $[\text{and}]^* = f^{\text{and}}$, that is: 'and' expresses a partial function which maps tuples of properties or states of affairs onto properties or states of affairs, or is undefined. This is the basis for a formal language treatment in the sense that here, some formal language item B corresponding to natural language item B^* is assigned a semantic value which is assumed to *be* the semantic values of B^*. According to the weak reading, [and]* merely *determines* such a function as follows:

> *Necessarily, there is a function* f^{and} *such that,* for any expression of the form 'being F and G', $f^{and}{}_{<F, G>}$ = *the property of being F and G, or* $f^{and}{}_{(<F, G>)}$ *is undefined.*

A property structure an expression E gives access to may contain some of the functions just described. Note that in addition to expressions determining functions and property designators, property structure terms may involve individual constants. For example, '_exemplifies the property of being loved by Peter' contains 'Peter'. They may also involve singular term forming operators. I shall ignore these expressions here, because they would make things unnecessarily complex – the account proposed here is limited to cases of property- and states of affairs-reduction. An interpretation of how individuals relate to properties they uniquely instantiate in terms of satisfaction allows for an expansion of the account (see Sect. 5.4.3).

We are now in a position to move from a given definiens, in the appropriate format containing expressions that explicitly designate properties, to terms that designate these property structures.

---

[7]If we were to introduce an exemplification-relation, then we could model this as follows: The exemplification relation behaves in a similar way: it takes properties and individuals or bound variables as arguments and maps them onto new properties or onto states of affairs (in case the number of individuals or bound variables is identical to the arity of the property).

5.3.3 An Operator for Designating Property Structures

Applying the following operator: 'The property structure of '_'' to expressions that are gained by the procedure just outlined from a maximally explicit characterization of the meaning of an expression *E*, we arrive at an expression, which transparently gives access to the property structure *E* give access to. Application of this operator forms the second step in our procedure to obtain property structure terms. So, this is how the operator functions:

It is a meta-linguistic operator. Its content can intuitively be captured by 'The way the (alleged)[8] designatum/significatum of '_' *is* according to '_'' (in both argument-positions, we have to substitute the same expression). PSs can thus be characterized as ways things are according to an expression or a description or a conceptual content. The operator builds upon the idea that to be a certain way according to a description is to instantiate certain properties according to a description, or to stand in a specific relation to other objects that exemplify specific properties according to the description.

This operator thus bears striking similarities to so called *story operators*,[9] although these operators are normally cashed out in terms of truths according to a story; they are sentential operators. In contrast, nothing is true according to the descriptive content of a singular term, or according to a rigid designator, or a predicate, just because these descriptive contents are not descriptive contents of sentences. However, we can introduce a similar idea in the present case. The idea of *being a certain way according to a description* might be further explicated as follows, given that an object is a certain way only according to a sentence or a proposition, rather than according to a predicate or a term: *an object x is F according to a description D iff according to the proposition that there is a y, which is adequately described by D, y is F, & y = x.* Thus, we now have a *sentence* ('there is an object which is adequately described by *D*') that expresses a proposition according to which an object is a certain way. If this object exists, then there is an object that *is* a certain way according to a description. Thus, the operator could be cashed out in terms of a sentential operator. What an object is according to a description can be known on a priori grounds, or on reflection on meaning alone; the content of the description gives the way an object is according to that description, and to be in a position to know that is to grasp what the description *means*.

Accordingly, water is, according to the description 'H_2O', an entity that stands in the instantiation relation to entities that stand in the part whole relation to three atoms, two of which exemplify the property of being hydrogen and one of which exemplifies the property of being oxygen. (Thus, sometimes an entity is presented by a description that is given in terms of *standing in a relation to*

---

[8]We need to be cautious; there might be cases where no object is designated at all.

[9]These play a role in models of make-believe (Walton 1990) and, without reference to the psychological aspect of make-believe, in Künne (1983, 1990).

objects (here: instantiation), which, in turn, stand *in a relation to objects* (here: part-whole relation), which, in turn, *are a specific way* (here: being hydrogen and being oxygen)). Thus, we have characterized the idea of a property structure term (‘PST’). *It is the result of applying, first, the procedure described above to a maximally explicit meaning characterization of a term E, and then applying the operator ‘The property structure of ‘_’’ to the result of the previous step.* We thus obtain an expression that designates the property structure of the expression we started with in an interesting, namely, a *transparent* way. This is a *Property Structure Term* (henceforth: PST). We have, thus far, primarily commented upon expressions designating PSs. Let us now briefly comment upon the metaphysics of PSs.

5.4 The Metaphysics and the Functioning of Property Structures

I suggest treating PSs as *tuples* of properties, functions, and tuples thereof. This enables us to intuitively model the semantic impact of the syntactic structure of PSTs in a convenient way.

The property structure ‘the property of being red and green’ gives access to is the tuple <and, <being red, being green>>,[10] where ‘and’ is treated as designating the partial function it determines in its ordinary use as described above. Similarly, ‘Peter loves Mary’ can be interpreted as <the loving relation, <Peter, Mary>>. Similarly, the property structure ‘_is a tiger or is red and green’ gives access to is this structure: <or, <<being a tiger>, <and, <<being red>, <being green>>>>. Thereby, we arrive at structured entities that clearly deserve the name of a property structure: Their constituents are properties or tuples of properties, facts (intuitively: an n-ary property whose positions are occupied by n objects), and functions. A more detailed account of how these property structures can be conceived of is proposed in the appendix. This requires a semi-formal treatment of PSTs; for example, to handle the scope of quantifiers that bind variables, we have to deviate from the grammatical surface structure of ordinary language PSTs. However, given this intuitive outline, we are in a position to account for several relations that are of crucial importance to explicate the notion of a constitutive structure. To repeat: *The property structure an expression E gives access to is a tuple of properties and, sometimes, functions on properties and states; it is a way an object is according to E.* As a limiting case, we might allow for property structures which just are one property: If ‘water’ lacks

[10]This already points to one problem: Tuples are finer individuated than one would normally assume the semantic value of expressions like ‘being red and being green’ to be. There is a difference between <and, <being red, being green>> and <and, <being green, being red>>. But is there a semantic difference between ‘being red and being green’ and ‘being green and being red’?

descriptive content, then the PS it gives access to is just *being water*, rather than <being water>. Building on this characterization, we can introduce the relevant notion of a constitutive structure and related notions.

5.4.1 Basic and Complex Constituents

It will be convenient to introduce the notion of a basic constituent of a property structure:

(Def. Basic constituents): *A basic constituent y of a property structure PS is a property referred to by some simple property designator (obtained from a simple predicate) or a kind term in a PST.*

Thus, all and only those constituents of a property structure that are referred to by a relevantly semantically basic constituent of a PST are basic. Now, we can also define the notion of a *complex constituent*. Complex constituents are, intuitively, tuples that may function as property structures themselves and that are not basic; they are a way an object is according to a semantically complex description of that object. For example, in the description 'being red and being green or being a lion', 'being red and green' designates a complex constituent of the property structure.

(Def. Complex Constituent): *A complex constituent CPS of a property structure PS is a non-basic constituent of PS.*

5.4.2 Determination of Objects

Property structures and their complex and basic constituents *determine* properties or states of affairs (recall that our property-structure talk does not allow for determination of, say, individuals) in a specific way:

(Def. Determination) *An entity x is determined by a property structure PS of an expression E iff x is the only object which is the way the referent or significatum of E is according to E.*

What is determined by a PS of an expression *E* will be identical to what *E* designates or signifies (if anything) iff the property structure an expression gives access to is similar to its meaning in the sense that it *determines its designatum* (as the Fregean would have it for Fregean sense). Note that sometimes, no object is that way – one may deny that there is the property of being a square triangle, although there might be a property structure that appears to determine a property. However, the proposition that *according to the proposition* that there is a property adequately described by the descriptive content of 'being a square triangle', there is a property of being a square triangle whose instances are squares and triangles does not entail that there is such a property. Thus, the existence of a property structure that seems to determine a unique entity does not ensure that the entity exists.

5.4.3 Satisfying Property Structures

Now, we are in a position to define a notion of satisfaction. Consider predicates and the PSs they give access to first. Take, for example, the predicate '_is the author of *The Life and Opinions of Tristram Shandy, Gentleman*'. The property so determined is the property of being the author of *The Life and Opinions of Tristram Shandy, Gentleman*. Now, this property *applies to* or is *satisfied by* the *x* which instantiates the property determined by this structure, namely the property of being the author of *The Life and Opinions of Tristram Shandy, Gentleman*. Thus, it is Laurence Sterne. This is so because the property of being the author of *The Life and Opinions of Tristram Shandy, Gentleman* is instantiated by Laurence Sterne only. Similarly, the property structure *being a horse* is satisfied by the set of horses, because the property determined by this structure is instantiated by the elements of this set. The property structure *the property of being the author of* is satisfied by pairs of objects which instantiate this property. Property structures which determine an n-ary relation (with $n > 1$) are satisfied by n-tuples of objects, namely, those tuples whose elements instantiate that relation. In the case of PSs which determine a unary property, the set of entities which instantiate the so determined property satisfies that property structure (or, if it is just one object, this object satisfies that structure). For property structures that determine a state of affairs, the object satisfying the property structure is the state determined by this property structure. We can characterize this in terms of an extension, where 'extension' is understood as follows: If a property structure PS determines a unary property *P*, its extension is the set of individuals that are P (or, if the property is satisfied by a one-set, we may allow as a convention to say that the extension = the only element of that set). If PS determines an n-ary property *P* (with $n > 1$), the extension is a set of n-tuples which instantiate P. If a property structure PS determines a state of affairs *F* then the extension just is this state of affairs.

(Def. Satisfaction) *An entity x satisfies a property structure PS iff x belongs to (or is) the extension of PS.*

On this basis, we can define the notion of a constitutive structure.[11]

[11]Based on these notions, we could describe the functioning of definite description forming operators, such as the jota-operator, when applied to a predicate that signifies a unary property with only one instance, as a function which takes what is determined by a PS to what satisfies the PS.

5.4.4 Constitutive Structures

The notion of a constitutive structure will enable us to account for what mimics the explanatory directionality at the conceptual level, or the level of property structures. Intuitively, the idea is that we can account for the directionality of reductive explanation as follows: the truth of a reductive explanation that *x* is *F* *because x* is *G depends* on facts about differences in how the property signified by '*F*' and '*G*' *is according to the respective descriptive contents* of '*F*' and '*G*'. This is parallel to the idea that a true causal explanation that *x* is *F because y* is *G* depends on, say, the laws governing, and metaphysical features of *x*'s being *F* and *y*'s being *G*; or, alternatively, that it depends on regularities among *F*s and *G*s. We thereby give information on what the causal explanation depends on. In the present case, I propose that reductive explanations depend on differences in property structures under which the reduced or reducing object is presented. And just like the accounts of mechanistic dependence discussed in Chap. 4 do not aim at a "reductive" account of mechanistic dependence in counterfactual or probabilistic terms, the present proposal does not aim at coming up with a "reductive" account of reduction. The difference in conceptual representation just is the difference in virtue of which the directionality of reductive explanation goes from water to H_2O, rather than the other way around.

We already defined a notion of reduction. This definition is, I submit, perfectly fine. However, we now enrich this definition by giving additional information on what sort of features reductive explanatory dependencies track. They do *not* track relational features of, or differences between reduced and reducing object. Rather, they track differences in modes of presentation, or differences in property structures, differences in the transparency with respect to the object's nature. We start with an indirect characterization. Reductive explanations track differences in property structures as follows: A property structure PS_1 is a constitutive structure of *x* with respect to a property structure PS_2 iff PS_1 and PS_2 determine the same entity, and whatever satisfies PS_2 does so because it satisfies PS_1. The property structure 'H_2O' gives access to is a constitutive structure of water with respect to the property structure 'water' gives access to because whatever is water, it is the way the concept of water has it (satisfies the property structure of 'water') because it is the way the concept of H_2O has it (satisfies the property structure of 'H2O'). We thus mirror (*Explication I*). Assume that 'water' is a directly referential term. Whatever satisfies the property of being water, it does so because it has the properties that correspond to the property structure 'H_2O' gives access to. Based on this indirect characterization of a constitutive structure, we can account for the candidate features of different property structures that back up the dependence relation, such as mereological relations: If reduction relations track mereological relations, then any property structure which presents us with an object in terms of its parts is at a "lower level" than any property structure which presents us with the same object as a whole, or at a "higher level" of composition. Recall the minimal condition on property-talk proposed above: properties are non-representational entities, entities in virtue of which causal objects behave the way they do. Being presented as having properties, that is: being presented by a property structure, can thus, intuitively, give access to the "causal architecture" of an entity. If a property structure gives access to the properties in virtue of which the object does what it does, or is what it is, it is a constitutive property structure. If an expression, such as 'H_2O' gives access to the causal organization of an object, then it is at a lower level than any directly referential expression which presents us with the same entity in a basic way, or than any expression that presents us with an object under a phenomenal property structure. If so, the property structure 'H_2O' gives access to is a constitutive structure with respect to the property structure 'water' gives access to. We can thus define a relational notion of a constitutive property structure.

(Def. Constitutive PS) *x is a constitutive property structure of an entity y with respect to a property structure x* iff x and x* are PS-s that both determine y, and are such that for every z, if z satisfies x* it satisfies x* because it satisfies x.*[12]

Again, the explanation is not the explanation of conceptual constitution. In this sense, the constitutive structure is *explanatorily prior* to the property structure with respect to which it is a constitutive structure. A non-constitutive property structure is a property structure that is not a constitutive structure with respect to any other property structure. A constitutive property structure is a property structure that is a constitutive structure with respect to some other property structure. How does this relate to reduction? 'Water reduces to H_2O' is true because (i) water $= H_2O$ and, given that directly referential expressions never give access to an entity's nature, (ii) the PS of 'water' is not a constitutive structure of water and the PS of 'H_2O' is a constitutive structure of water. This is the paradigm case of reduction of folk-theories to scientific theories. However, this structure is not preserved in inter-theory reductions. Sometimes, both, the *explanans* and *explanandum* in a reductive explanation as well as '*a*' and '*b*' in '*a* reduces to *b*' present the target under constitutive structures. A quantum mechanical description of water will, if physicalism is correct, be at a lower level than the description of water in terms of H_2O. Thus, we need to distinguish levels of constitutive structures. Fundamental constitutive structures are property structures which are constitutive, and which are such that there is no property structure which is constitutive with respect to it. But why stick to explanatory notions in the explication of a constitutive structure? Why not try to spell it out in terms of promising mereological relations? For one, a notion of reduction should be independent of such strong metaphysical commitments.

---

[12]Note that in case the property structure determines an individual (given an appropriate operator) or a fact, the object satisfying the structure is the object determined by the structure. I assume that the definition could easily be given in terms of determination, rather than in terms of satisfaction. One may want to discuss what is prior: explanatory relations among instances or explanatory relations among properties. Here, I chose to describe the notion in terms of a relation among instances, without being committed to any such priority claim.

As pointed out before, reduction is neutral with respect to what actually is the fundamental level. But this is not the only problem mereological interpretations of reduction face.

5.5 Constitutive Property Structures and Mereology

Let me first illustrate the hope to define the notion of a constitutive structure in a way that yields a "reductive" account of reduction, that is: explicate it independently of explanatory notions, assuming that reduction tracks mereological relations. It will turn out that even the naturalist should not describe reduction in mereological terminology, unlike, for example, Schaffner (1993) and Horst (2007) suggest.

The lower the level of a constitutive structure, the deeper the access we get to an object's nature. The set of hydrogen atoms satisfies the property structure *being hydrogen* (or an appropriate complex structure which is based on the meaning of 'hydrogen'). Similarly, the set of oxygen atoms satisfies the property structure *being oxygen*. In this sense, the property structure given by H_2O gives us water in terms of its (instances') parts. Put differently: The way water is according to 'H_2O' is a way its (instances') parts are. In this case, being a constitutive structure relates to mereological relations. This seems to support the assumption that reduction is tied to mereological relations, an assumption seemingly shared by Schaffner (1993) (see Sect. 7.4), Kim (2008) and others (Chalmers 1996, 43; Fodor 1974, 107). Consider the following case: Take, H_2O and a quantum mechanical description of water, here abbreviated '$water_{quantum}$'. Both give access to a constitutive structure of water, but nevertheless, we should assume that H_2O reduces to $water_{quantum}$, or that something is H_2O because it is $water_{quantum}$, and that, thus, we have to account for a relevant difference between the distinct constitutive property structures these terms give access to. What does this difference consist in? Again, one might feel tempted to tie the difference to mereology.

For cases where reduction tracks mereological relations, we can give a criterion for the directionality of reduction that is independent of explanatory notions. One constituent of water has the property of being an oxygen atom *because* it has certain physical features that are revealed in a quantum-mechanical description. Similarly, an appropriate answer to what an oxygen atom basically *is*, we may expect an answer in quantum-mechanical terms. And, similarly, an oxygen atom connects to two hydrogen atoms under specific conditions because of its physical features. Thus, the explanatory dependence can now be accounted for in terms of objects determined by (basic or complex) constituents of the property structure that depicts the object at a higher level of mereological composition. Accordingly, we can introduce a relativized criterion for level-comparison, that is based on how property structure mirror mereological constitution. Let me illustrate this point: The property structure a description of water that is a mix of deictic-expressions (that give us the property of having two constituents that are hydrogen-atoms) and a quantum-mechanical description of the oxygen atom gives access to is at a lower level than the PS 'H_2O' gives access to with respect to the property of being an oxygen-atom. A quantum mechanical description will contain properties of "parts" of O-molecules, that is: of objects that satisfy a constituent of the property structure 'H_2O' gives access to. These properties and parts are not reflected in the property structure 'H_2O' gives access to.[13] We thus get: *A constitutive structure CS is at a lower mereological level than a property structure PS with respect to some constituent P of PS iff CS and PS determine the same entity, and CS contains some complex structure C that determines the same entity P determines, and C, unlike P, consists of properties that are satisfied by parts of objects that satisfy P.* But does reduction track mereological dependence?

If pain reduces to C-fiber stimulation, then pain events are not given in terms of "parts" of pain. Ignore the issue of "parts" of events (an issue mechanistic accounts may shed light on); let pains be possible activation of *one* C-fiber; then, there is no proper part of the object of C-fiber activation (namely, the C-fiber) which plays a role in the reduction. Similarly, one might want to say that *being an iron constituent* reduces to *being a Fe-atom*. If so, we have reduction without mereological decomposition – Fe-atoms are not parts of the iron-constituents we aimed at; they just *are* these iron-constituents. Thus, part-whole relations are not always crucial for our understanding of reductions, even if naturalism is true.

In addition, let me repeat the more general reason to describe reductive dependence independent of mereological dependence, already discussed in Chap. 2. Assume that an idealist and a naturalist discuss issues of reduction. They agree upon everything except for the directionality of reduction. Assume that the idealist is not conceptually incoherent when claiming that, say, C-fiber firing reduces to a mental entity, and *that this mental entity does not have parts*. If so, we should not define our notion of reduction in mereological terms. Mereology does not come for free for the reductionist. It is a substantial point to claim that at least some things reduce to their parts. Thus, we should not explicate reduction in terms of mereological relations. Using the apparatus of property structures, we can now come up with a more thorough explication of reduction.

5.6 An Explication of Reduction

Possession of the notion of levels of constitutive structures enables us to give truth conditions for reduction statements of the form '*a* reduces to *b*', where the terms substituted for '*a*' and '*b*' do not designate theories.

---

[13]This also matches inter-level integration, as, for example, suggested and discussed in (Darden and Maull 1977; Schaffner 2006; Craver 2007). We may replace some descriptions by others, but do not fully reduce one level. Neuroscience makes use of behavioral notions in order to capture the behavior the neuronal patterns influence or are influenced by.

5.6.1 The Core Notion

The definition is given in a schematic way: For '*a*' and '*b*' substitute kind terms (or event-terms, or property designators . . .) or terms referring to states of affairs – we thus arrive an a general explication:

(Explication – Schematic) *A sentence of the form 'a reduces to b' expresses a truth iff*

(i) *a* = *b, and*
(ii) *'a' gives access to property structure PS^A, and 'b' gives access to PS^B*
(iii) *PS^B is a constitutive structure of b with respect to PS^A.*

There are two ways for PS^B to be at a lower level than PS^A: either the latter is not a constitutive structure, or it is a constitutive structure which is at a higher level. The second and the third requirement model the relevant meaning aspect. This definition makes transparent that reduction is not asymmetric. This does not mean that it is symmetric – recall the discussion in Chap. 3. We can also give a non-schematic definition, building on the idea that sensitivity to semantic facts other than reference or designation is captured by formulations such as '_is presented under/as being _':

(Explication – non-Schematic) *x, when presented under PS1, reduces to x, when presented under PS2 iff*
PS2 is a constitutive structure of x with respect to PS1.

This is supposed to capture a generalized version of the idea underlying (*Explication I*): that *F-ness* reduces to *G-ness* if and only if (i) for every *x*, if *x* is *F* then (*x* is *F because x* is *G*), and (ii) *F-ness* = *G-ness*. It captures this idea, but it is not limited to this notion of reduction. To illustrate this point, consider, for example, a case of what one might describe as fact-reduction: The fact that this amount of water freezes reduces to the fact that this sum of H_2O molecules forms lattice structures. Here, what is presented under the property structure of a sentence is a *fact* (a worldly entity, such as a Russellian proposition).[14] We arrive at the following derivative characterization, which is a special case of reduction as covered in the definition just proposed:

(Explication III – Facts) The fact that *p* reduces to the fact that *q iff*

(i) *p* because *q*, and
(ii) the fact that *p* = the fact that *q*.

[14]This idea is inspired by the suggestion of an anonymous referee; according to this suggestion, we could describe theory reduction in terms of fact reduction – the fact that the reduced theory is true reduces to the fact that the reducing theory is true. This idea is picked up again in Chap. 8, footnote 4.

Further derivative explications will be proposed below. It is now time to turn to a first evaluation of the proposed explication.

An expression presents an object either under a concept or does not present it under a concept (if it is a directly referential expression). An object, if presented under a concept, is presented as having certain properties. Different concepts present objects as having different properties. Sometimes, an object instantiates some properties because it instantiates other properties. If materialism is correct then water has the property of being tasteless partly because it has the property of being constituted by H_2O-molecules. This connection is made transparent in reduction-statements that connect terms that express a conceptual content. For reduction statements in which the term on the reducing side lacks conceptual content, things are different: Here, the concept expressed by the term on the reducing side is transparent to at least some properties of the object in virtue of which it behaves the way it does, or is what it is. Dependencies between an object's satisfying different property-structures captures this very idea. But how to argue for the idea that this model is promising? First, we can draw an intermediate conclusion: To the extent that this definition is *sound* and *captures the relevant intuitions concerning the notion of reduction*, it shows that skepticism about notions of ontological reduction is misguided. We can give an appropriate explication. Hence, we should not be afraid of this kind of reduction.

Recall the job-description given above:

(Job-Description): We need a definition that

(i) accounts for the directionality, such that it gives a definition according to which if a reduces to b, then necessarily, b does not reduce to a, and
(ii) accounts for the idea of unity (ideally in the sense of strong unity) without elimination.

In addition, it will be desirable that our definition enables us to

(iii) illuminate the paradigmatic cases;
(iv) explain the intuitive characterizations of the notion of reduction;
(v) account for related topics, such as reduction and physicalism, reduction and scientific unification, and similar issues; i.e. the explication should yield a fruitful notion by fulfilling the explanatory task associated with 'reduction'.

We have shown that the explication meets criteria (i) and (ii). Let us briefly apply the explication to paradigmatic cases and informal characterizations of reduction in the literature, showing that it also meets criteria (iii) and (iv). The model perfectly matches the guiding intuitions. In the next chapters, I will argue that it is, in addition, fruitful, and, hence, meets criterion (v): It can be applied to the reduction debates in the philosophy of mind and the philosophy of science as well as to a number of related, unsolved problems tied to reduction, such as the epistemology of reductions, reductive explanations and interventions, and reduction and unification.

The explication captures the paradigmatic cases. We have already commented upon the water-H_2O case. Let me now comment upon the two cases Nagel mentions: mean kinetic energy and temperature, and headaches and what gives rise to them, to illustrate the idea. The meaning of 'mean kinetic energy' gives access to a property structure which is at a lower level than the property structure the meaning of 'temperature' gives access to – it gives the resources to explain temperature in terms of events some object's entities engage in, which give rise to, and, in this sense, *reductively explain* the occurrence of temperature. Having a specific temperature is, by assumption, having a specific mean kinetic energy. Moreover, parts of objects that have a certain temperature must behave in a specific way for the occurrence of a specific temperature. Finally, this behavior of parts is captured by the property structure under which 'mean kinetic energy' presents us with temperature. Thus, the case is covered by the idea of a constitutive structure. Note again that this is not very surprising: The notion of a constitutive structure is an explanatory notion. It just reveals the explanatorily relevant aspects of the concept under which 'mean kinetic energy' presents us with temperature.

Similarly for headaches: The conditions for the occurrence of headaches are here interpreted as being fixed by a constitutive structure the basic properties of which belong to a lower level. Some constitutive structure could fix "the detailed physical, chemical, and physiological conditions for the occurrence of headaches" (or one of these), such that once the relevant constitutive structure is "ascertained [...] an explanation will have been found for the occurrence of headaches." (Nagel 1961, 366) Thereby, we give an interpretation of what Nagel seemed to have had in mind introducing these examples to illuminate the notion of reduction.

The same holds for (the internal process of) social cognition and the mirror neuron mechanism, pain and C-fiber stimulation, and water and H_2O. Each of these examples comprises a re-description in terms of constitutive structures, either in terms of parts or merely in terms of properties that are explanatorily relevant for the occurrence of properties under which the object is presented by the higher-level description. Thus, one condition on an appropriate explication of the notion of reduction is met: It should cover the relevant examples. It does not only cover these examples, it also *explains* why they are cases of reductions (given that the relevant identity links actually hold true): They form correct reductive statements because the target of the *explanandum* is given by a property structure in the *explanandum* which is *not constitutive*, and in the *explanans* it is given in terms of a constitutive property structure. If it were constitutive in the *explanandum*, it had to be constitutive *at a lower level* in the *explanans*. Thus, the explication of the notion of reduction not only covers, but also illuminates the paradigmatic cases of alleged reductions.

The intuitive characterizations we find in the literature are also covered. The explication of reduction can be used to illuminate the somewhat metaphorical or at least underdetermined sketches of what reduction consists in, sketches that build upon an intuitive understanding of being *over and above*, of *levels*, of *assimilation of distinctive traits,* and of *macroscopic phenomena*. Let me list these statements again:

(Kim)	If Xs are reduced, or reducible, to Ys, there are no Xs over and above Ys. (cf. Kim 2006, 275f., (given in a similar fashion by Smart 1959))
(Wimsatt)	Inter-level reductions are compositional. They localize, identify, and articulate mechanisms that explain upper level phenomena, relationships and entities. (Wimsatt 2006, 449)
(Chalmers)	[W]hen [in the context of a reductive explanation, *RvR*] we give an appropriate account of lower-level processes, an explanation of the higher-level phenomenon falls out. (Chalmers 1996, 42)
(Nagel-1)	[In reductions, A] set of distinctive traits of some subject matter is assimilated to what is patently a set of quite dissimilar traits. (Nagel 1961, 339f.)
(Nagel-2)	[The reduced] science deals with macroscopic phenomena, while the [reducing] science postulates a microscopic constitution for those macroscopic processes. (Nagel 1961, 340)
(Sarkar)	The reduced theory is explained by a reducing theory which is presumed to be more fundamental. (Sarkar 1992, 167)

For the Kim/Smart-intuition, we get a straightforward re-interpretation: Reduction guarantees that the reduced entity does not exist over and above the reducing entity because it requires *identity* of the two. Wimsatt's assumption and Nagel's second characterization are covered and explicated in a similarly straightforward way: The notion of a *constitutive property structure* gives a precise idea of the notion of a "microscopic constitution" (Nagel), and it matches the idea that in reductions, we "identify, localize and articulate" (Wimsatt) the composition of mechanisms or, more broadly, entities when presented as being constituted in a certain way; although reduction does not essentially track mereological dependence, it tracks mereological dependence where mereological dependence matches reductive dependence. The explanatory aspect pertinent in Wimsatt, Chalmers, and Sarkar is built into the model of reduction proposed here. Now, consider Nagel's first remark: That a "set of distinctive traits" is "assimilated to what is patently a set of quite dissimilar traits". This is achieved by identifying an entity presented under one specific property structure with an entity presented under one *dissimilar* property structure. Conceive of distinct sets of traits in terms of distinct property structures. These are clearly dissimilar. Despite the fact that they are dissimilar, like the property structure of 'water' and that of 'H_2O', they might nevertheless determine the same entity. Now, Nagel seems to think of more general relations, not only covering *two* property structures which determine one and the same entity, but rather covering pairs of sets of traits which are the unique sets of traits associated with two theories or sciences. Assume that folk-chemistry gives us chemical kinds under rigid designators, or under phenomenal properties. Chemistry gives us chemical kinds in different ways. Then, once we are able to show how everything folk-chemistry gives access to can be reduced to chemistry, we have really shown what Nagel seems to have had in mind: The set of properties under which chemical entities are presented by property structures the language of folk-chemistry gives access to is thus assimilated to the set of properties under which chemical language gives access to the same entities.

Similarly for psychology and neuroscience: We may try to reduce psychological traits (the properties under which mental kinds are presented in folk-psychology) to neuroscientific traits (the properties under which mental kinds are presented in neuroscience). Once we succeeded, these dissimilar traits are *assimilated.* Thereby, it is shown how these traits connect (in Chap. 9 this topic and its connection to unification and scientific levels will be addressed in more detail).[15]

Thus, the constraints imposed on an appropriate explication of the notion of reduction by the intuitive descriptions are met. If these intuitive sketches of reduction are representative, it has been shown that the concept of reduction as explicated here does not violate the constraints imposed on it by its use in philosophical discourse. Moreover, we can introduce fruitful explications of derivative notions, building on the framework just proposed.

5.6.2 Derivative Notions of Reduction: Partial Reduction, Plural Reduction, and Generic Reduction-Statements

Building on the core notion, we will later be able to define notions of token-reduction, type reduction and theory reduction. Three notions of reduction that might be of particular importance, and which are not so easily obtainable from the above definition, are those of *partial reduction*, where the reduction base is not fully represented in the explanans of a reductive explanation, of *plural reduction*, where in a reduction statement, the explanans refers to a conjunction of constituents of the target of the explanandum (an idea pertinent in the debate on mechanistic explanation (see Chap. 4)), and what one may want to call a 'generic' notion of reduction (or better: 'generic reduction statements'), which is best re-defined in terms of a meta-semantic notion. Intuitively, it is this notion that enables us to express general theses about how particular reduction statements are to be shaped. Partial reduction is parasitic on a notion of plural reduction. So, let us first address this latter conception.

5.6.2.1 Plural Reduction

How to account for cases such as 'the heart pumps blood by the organized contraction of muscle fibers' within a reductive framework? Is there no option for the reductionist to accommodate such cases, without buying into a theory of non-derivative plural identity? There is. Recall the brief discussion of reduction and

[15] The idea of assimilation thus makes it hard to conceive of reductions to come in a case by case manner. We do not reduce particular traits and then conclude that we can reduce the entirety of, say, chemical traits. So, it is important to note again that here, I am not concerned with explaining how reductions are carried out.

mereological dependence (see Sect. 5.5). Assume that mereological dependence is one form of reductive dependence. Then any sum reduces to its (proper) parts. Take, as an example, 'this statue reduces to this particular atom, and to this particular atom, and . . . ' (and let us grant that statues are nothing but the material they are made of). Does it follow that this statue = this particular atom, and this particular atom, and . . . ? Here is a suggestion how to bypass the topic of plural identity in the context of reduction. There is an easy operation on the expression on the right hand side of the identity sign, which generates a singular term; namely, applying 'the mereological sum of_' to this expression. We thus arrive at a truth, and an interesting one: The sentence 'the statue reduces to the mereological sum of this atom, and this atom, and . . . ' can be treated in the ordinary way. If the reduction statement is true, then this particular statue is this particular statue *because* it is the mereological sum of this atom, and this atom etc., and this particular statue is identical to the mereological sum of these atoms. How can we exploit this idea for other cases, which do not involve mereological dependence?

If plural reduction statements of the form '*a* reduces to *b1, b2* . . . ' are true, then there is a relation *R*, such that there is a unique *x* which is such that *x* stands in *R* to *b1, b2* . . . Sums of objects are the unique entities which stand in the *being the mereological sum of*-relation to their parts. For other cases, such as complex events, the reductionist should assume that there is a similar relation. Otherwise, we do not get the reduction: in a straightforward sense, the whole or composite would be more than its parts. Thus, here is a suggestion for plural reduction:

(Explication IV – Plural -Reduction) *x, when presented under PS1,* ***plural-reduces*** *to* $y_1, \ldots y_n$ *iff*

(i) *there is an object z, a relation R such that* $z\ R\ y_1 \ldots y_n$, *and there is a property structure PS2, and*
(ii) *PS2 describes z as standing in R to* $y_1, \ldots y_n$, and
(iii) *x, when presented under PS1, reduces to z, when presented under PS2.*

Thus, cases of plural reduction can be accounted for within this framework without commitment to plural identities. Based on this notion, we can easily explicate a notion of partial reduction.

5.6.2.2 Partial Reduction

A partial reduction just gives an incomplete reduction base, but it does so in the right terminology. The notion of a partial reduction is a technical notion that is supposed to mirror mistakes or blind spots in alleged reductions. For example, even if mirror neurons play an important role in the cognition of others as social beings, this does not mean that they do the relevant job alone, and it seems rather far fetched to claim that the cognition of others as social is nothing but the activation of certain patterns of mirror neurons. Being the activation of the relevant pattern of mirror neurons may thus be an essential part of the process of social cognition, but the property structure this description gives access to does not determine social cognition. The suggestion I wish to make is based on the notion of plural reduction just defined. Intuitively, a partial reduction is a plural reduction where the list of objects that together form the reduced entity is not complete – this is probably standard in mechanistic explanations, so that mechanistic explanations can be conceived of as partial reductions, if monism is true.[16] Interesting partial reductions will be partial reductions with an eye on especially interesting aspects of the reduction base. Social cognition in humans may involve processing of visual information. This is an interesting topic, but it is relatively well understood. Thus, focusing on the mirror neuron mechanism, a particularly relevant part of the reduction base is described. So, here is the explication.

(Explication V – Partial Reduction)	*x, when presented under PS1,* ***partially reduces*** *to y1, yn ... iff* *there are some objects y2, ... ym, (with {y1, ... yn} ∩ {y2, ... ym} = ∅) such that x plural reduces to y1 ... yn, y2 ... ym.*

Correspondingly, we may label the core notion of reduction defined above 'full reduction'. Let us now turn to constructions in which 'reduction' is frequently used, which express general theses about reduction relations.

5.6.2.3 A Generic Notion of Reduction

Consider the sentence: 'This table is *nothing but* (the sum of) this tabletop, these legs and these screws'. Although the 'nothing but'-locution is symmetrical, it may be used to articulate one's reductionist inclinations; one may wish to continue: '... this table *reduces to* (the sum of) this tabletop, these legs and these screws'. Such sentences may motivate the generic claim that *wholes are nothing but, and, hence, reduce to (sums of) their parts*. In quantified sentences, we can use the reduction predicate. In these contexts, it does not perfectly fit the description proposed above. Although one may wish to claim that *being a whole* (fully) reduces to *being a sum of parts* in the sense of the explication proposed here, one may want to state that wholes (fully) reduce to (sums of) their parts in order to articulate the generic statement that whatever is a whole, it reduces to its parts. The latter claim is not adequately captured by the former. To see this, consider the claim that tables (fully) reduce to (sums of) their parts. This is not to be conflated with the claim that *being*

[16]The use of 'partial reduction' thus deviates from the use employed by Schaffner (2006, 2012), who takes partial reductions to be less than fully fledged Nagelian theory reductions, or, alternatively, something close to local (though possibly full) mechanistic explanations. A number of the reductions covered by the proposal offered here as *full reductions* would, under Schaffner's notion of a partial reduction, probably turn out to be partial reductions. See Sect. 7.6 for a discussion.

a table (fully) reduces to *being a sum of parts*. Another prominent example is the generic claim that mental properties (fully) reduce to physical properties. So, how to give an explication of this phenomenon? (Although this may not be the appropriate way to handle generic statements, I talk as if there were a generic *notion* of reduction, so that the concept expressed by the predicate in these contexts would turn out to be responsible for the phenomenon, rather than the entire sentence.) I suggest explicating generic statements as meta-linguistic claims about conditions on appropriate modes of presentations in true reduction statements. On this view, mental properties reduce to physical properties if and only if for every *x*, if *x* is a mental property, then there is a mode of presentation *m1*, there is a mode of presentation *m2*, such *x*, when presented under *m1* reduces to *x* when presented under *m2*, and *m1* presents us with *x* as mental, and *m2* presents us with *x* as physical. This explication hinges on the assumption that modes of presentations, and, hence, property structures can present us with objects as mental, or as physical, or as wholes, or as sums of parts. This idea will be made explicit in Sect. 9.1. Until then, we will have to rely on an intuitive understanding of property structures that present us with objects as being of a certain kind. Let us, for cases like 'wholes reduce to their parts' and 'mental properties reduce to physical properties', express this idea saying that a property structure presents us with what it determines as a mental object *if and only if* it is a mental property structure.

(Explication VI – Generic -Reduction) *F-s reduce to G-s iff for every x that is an F, there is a PS1, a PS2, such that*

(i) *x, when, presented under PS1 reduces to x, when presented under PS2, and*
(ii) *PS1 is an F-property structure, and*
(iii) *PS2 is a G-property structure.*

For the claim that *tables* reduce to their parts, we have, ignoring the more complex quantification involved, to make a little detour: if tables reduce to their parts then they do so in virtue of the fact that there is a type of property structures to which the property structure 'table' gives access to belongs. This type, say, being a material whole,[17] is such that its instances, when presented under property structures that present us with them as wholes or artifacts, reduce to themselves when presented under property structures of a different type: namely, under property structures which present us with them as being constituted by parts (and their properties) of such wholes. This aspect is already pertinent in the above explications of plural

[17]Note that it is not clear at all to which type a given property structure belongs; most of them will belong to more than just one such type. Which type is actually at stake (or which, in an appropriate reconstruction of such statements with the goal of a fruitful explication) will probably be determined by context: In a debate about material objects, where participants agree that all material wholes have the same metaphysical status, but do not agree as to what this status *is*, we should not count reference to tables as giving an example of a property structure of the type artifact-property structure. In a debate about the status of artifacts, however, this may be appropriate.

and partial reduction: on the right-hand side of the reduction predicate, no specific property structure is required, under which the plurality of objects is to be presented. The discussion of generic reduction suggests that this is so because mentioning of the objects is sufficient to point to the fact that it is in virtue of being constituted by these objects, and in virtue of these objects' properties, that the reduced item behaves the way it does, or is what it is.

The account presented here thus covers the core uses of 'reduces to'. However, it builds upon a specific interpretation of the relevant relational features of the expressions that pick out the reduced entity in true reduction statements. Other accounts may build on other candidate features. Two such possible proposals, a semantic and a pragmatic one, will briefly be discussed in the next section.

5.7 Rival Explications

As laid out in Chap. 2, the proposed characterization has the status of an explication. An explication derives its merits and deficiencies from various aspects. Partly, its quality hinges on the vocabulary it employs. As suggested in Chap. 2, two explications may yield strikingly similar results despite the fact that they exploit vastly different conceptual resources, as long as they exhibit similar structural features. This section is supposed to contrast the present proposal with two rival interpretations one may come up with in order to account for reduction; they are both compatible with (Explication I), but suggest different interpretations of the source of the directionality. These rival interpretations are structurally similar to the one proposed here, but they do not carve reduction in terms of *property-structures*. The first uses a two-dimensional framework, the second is more pragmatist in spirit. I will leave it at intuitive sketches, so as to make clear how the apparatus presented above would have to be adjusted in order to fit these rival accounts, and to indicate why I think that carving the explication in terms of property structures is advantageous.

5.7.1 Reduction Within a 2-D Framework

David Chalmers' attempt to model core features of Fregean senses within a two-dimensional framework provides the resources to mimic at least some of the features presented above, so that a similar explication in terms of the 2-D framework can be given. On this view, what has here been called a 'conceptual' difference, turns out to be an epistemological difference – a difference in epistemically defined primary intensions of expressions. Let me, first, introduce the core ideas of Chalmers' theory, then apply it to the case of reduction and finally hint to two problems that seem to occur within the framework. The summary to follow draws heavily on my (van Riel 2011).[18] According to two-dimensionalism, an expression can be associated with two semantic values – a primary intension and a secondary intension, which, together, generate a third value: a two-dimensional intension. Chalmers interprets primary intensions as functions from scenarios to extensions at scenarios and secondary intensions as functions from possible worlds to extensions at possible worlds. Chalmers intends to use this framework in order to model a number of relevant aspects of meaning, and to give an idea of how a priority, necessity, and conceptual analysis interconnect. Primary intensions are used to model something like Fregean sense. In order to understand the notion of a primary intension, we have to explain the notion of a *scenario* in the first place, and we have to explain how this notion is used in order to better grasp how meaning, or Fregean sense, relates to cognitive significance. There are two ways to conceive of scenarios: *scenarios as centered worlds* and *scenarios as epistemic possibilities.*

According to the centered world interpretation, scenarios are triples of ordinary possible worlds, individuals, and times. Individuals and times are introduced to handle indexical claims; they fix a certain point of view on or in the world. A primary intension of a sentence *S* obeys the following principle: *S*'s primary intension delivers the value TRUE for a scenario *W* if and only if the conditional 'If *W* is actual, then *S*' is a priori. Here, we consider a world as actual. Scenarios are associated with canonical descriptions. These canonical descriptions are supposed to guarantee that the material conditional of the form 'if *W* is actual, then *S*' comes in an appropriate form. There might be many different ways to refer to *W* we can use to generate instances of 'if *W* is actual, then *S*', that do not suit the purpose of evaluating *S* at *W* using a material conditional of this form.[19] So, in a sense, it is the appropriate descriptions of worlds that do the job of fixing the criteria for a primary intension, not the worlds themselves. To be more precise, the relevant conditional should be described as follows: 'If *W* (under description *D*), is actual, then *S*'. On different occasions, Chalmers proposes different ways one might conceive of primary intensions, ways that do not involve the notion of apriority. They all have in common that they use epistemic notions to evaluate the relation between a scenario under a description and a given expression; these differences do not matter for what follows.

According to the second alternative, scenarios are described in epistemological vocabulary. According to this interpretation, we should conceive of a scenario as a complete description of a way the world might be, which meets two requirements: It is epistemologically possible, and it is complete, such that there is no question which cannot be settled with respect to this description on a priori grounds (Chalmers 2004). More precisely: In this case, a scenario is an *equivalence class* of sentences of (non-natural) language *L*, which is such that (i) any member, *D*, of the class is epistemologically possible, and which (ii) is complete in the sense that no

---

[18]For a detailed introduction, see (Chalmers 2004).

[19]For a detailed discussion of the notion of a canonical description and different ways to interpret or replace apriority, see (Chalmers 2004, 2006).

sentence of L is indeterminate given D. *Epistemic possibility* is cashed out in terms of apriority or similar notions: A proposition that p is epistemologically possible if and only of it is not a priori that not p, and that p is epistemologically necessary iff that p is a priori. Intuitively, primary intensions, on this interpretation, behave similar to how they behave in the scenarios-as-centered-worlds case: Consider a description D of the equivalence class. Then, for a sentence S, its primary intension yields TRUE for a scenario under D if and only if 'If D, then S' is a priori.

Up to now, we have focused on sentences. Primary intensions for sub-sentential expressions pose special problems, so I will merely comment on primary intensions of singular terms that do not involve indexicals, and I will presuppose that we take scenarios to be centered worlds. In this relatively simple case, a singular term's ('S') primary intension is a function from scenarios (W) under descriptions (D) to extensions (E), such that the relevant instance of the schema "If W (under D) is actual, then $E = S$" is a priori. For example, if a world in which the brightest star in the morning is Riegel Delta were actual then Hesperus would be Riegel Delta. This inference is a priori/based on purely rational grounds (this gives an idea of what it is to consider a world *as actual* to account for an expression's intension).

Whereas the first approach to scenarios is based on the notion of metaphysically possible worlds we are familiar with, the second approach describes scenarios in epistemological terms from the beginning. In order to construct primary intensions, epistemological notions are required in both cases. The basic idea is that primary intensions are functions that obey epistemological constraints. I will use the term 'scenario' in both cases, just like Chalmers does. Given that primary intensions can be conceived of as *structured entities* (see Chalmers 2011), and that primary intensions are supposed to capture Fregean senses, or modes of presentation, we could define the core notion of reduction in a way strikingly similar to the way the notion was explicated above; a (possibly structured) primary intension we arrive at after we analyzed the conceptual content of an expression may be more or less transparent with respect to the nature of the object it designates or signifies. If so, we may arrive at a hierarchy of primary intensions for objects, a hierarchy that mirrors a reductive hierarchy in a way strikingly similar to the way property structures mirror such a hierarchy. Then, intuitively, 'a reduces to b' expresses a truth iff a, when presented under the primary intension of 'a', reduces to a, when presented under the primary intension of 'b'. This will be the case iff $a = b$ and the primary intension of 'b' is transparent with respect to the nature of b, and the primary intension of 'a' is not transparent with respect to the nature of a, or it is less transparent with respect to the nature of b than the primary intension of 'b'. For primary intensions that determine properties, we can easily illustrate how we arrive at corresponding notions of determination and satisfaction: An object or a set of objects satisfies a primary intension iff it is the unique object or set of objects that instantiates the property determined by this intension. Given that properties are secondary intensions (let us assume that for the sake of simplicity), then the property determined by the primary intension of a property-designator 'a' just is the secondary intension of 'a'.

Note firstly that there is an interesting connection between this characterization of reduction and the characterization proposed above: Assume that every primary intension (at least every primary intension that occurs in a simple or complex descriptive content of an intension that actually picks out an object) determines, in the ordinary sense, the property of being the way the primary intension would have it (maybe the secondary intension). The (non-structured) primary intension of '_is tasteless' determines *the property of being tasteless*. If this generalizes, and if we treat primary intensions of logical vocabulary as specifying functions on primary intensions and corresponding functions on properties, and if the explication within the 2-D framework is correct, then the explication proposed above is correct. For any structure of primary intensions, we will get a property structure of corresponding properties.

So, can we decide between the two? The proposal offered above is less committal in the sense that it does not commit one to a particular notion of Fregean senses, like Chalmer's 2-D framework does. If sparse commitment translates into quality, the framework proposed above is advantageous, especially since everyone who endorses 2-D semantics is free to endorse it, or a 2-D version of it, as well. (I guess that similar arguments work for algorithm-based, or procedural semantic theories of Fregean senses.)

But there is more to be said: primary intensions are described in purely epistemic vocabulary. This feature is, in the context of explicating a notion of reduction, highly problematic, because it blocks a way to explain the cognitive or epistemic differences between reduced and reducing item. Since pragmatic and alternative "cognitive" accounts suffer from a structurally similar problem, I will turn to these first and then close with a critical note on this point.

5.7.2 Reduction Within Pragmatic and Cognitive Frameworks

Another obvious candidate to illuminate the relevant aspect that is responsible for the directionality is this: The difference is *cognitive and pragmatic* in nature. This idea has been put forward, for example, by Robert Van Gulick in a series of papers (1992, 2001):

> [T]he problem is to find a way, if possible, for us to use the contextually embedded resources of the reducing theory to do the equally contextual representational work done by the items in the theory we are trying to reduce. Nor should we ignore the [...] pragmatic parameter [...]. Success in real-world representation is in large part a practical matter of whether and how fully our attempted representation provides us with practical causal and epistemic access to our intended representational target. A good theory or model succeeds as a representation if it affords us reliable avenues for predicting, manipulating, and causally interacting with the items it aims to represent. (Van Gulick 2001, 14)

From this characterization, that is offered with an eye on the *pragmatic and epistemic success* of reductions, we can easily derive a characterization of how we could reconcile diversity and directionality with unity: '*a* reduces to *b*' expresses a truth only if $a = b$ and there is a difference in pragmatic and representational features of '*a*' and '*b*' (where representational or conceptual features are described as cognitive features).

A similar idea, cast in purely epistemic terminology, seems to underlie a proposal offered by Berent Enç (1976). On Van Gulick's view, conceptual differences are ultimately to be explained in cognitive and pragmatic terms. In contrast, Enç, employs the term 'epistemological framework', an expression he does not explicate in detail. His description of the situation comes close to the puzzle discussed in Chap. 3: According to Enç, there is a generative relation that reconciles explanatory directionality with identity. This generative relation is dependent on differences in "epistemic frameworks". Unfortunately, this notion invokes an unexplained notion of causation, and it does not give the resources to see how the puzzle vanishes; rather, the suggestion mainly consists in accepting that there is a tension in the assumption that identity in reduction may go together with what Enç describes as explanatory or causal asymmetry. He writes:

> The assumption [that an explanatory asymmetry is sufficient to exclude identity] is false. Explanation is an epistemological endeavor. Identity is a metaphysical fact. It is entirely possible for two property descriptions 'φ' and 'ψ' to refer to the same property in all possible worlds, and yet, since we may know that an object answers the description 'φ' antecedently, a new theory that shows the object to answer the description 'ψ' may succeed in telling us why the object answers the first description, or it may succeed in deepening our understanding of the fact that the object answers the first description. (Enç 1976, 290)

The 'epistemological' features of explanations Enç refers to are left unexplained. Again, on this view, '*a* reduces to *b*' expresses a truth only if $a = b$ and '*a*' and '*b*' differ with respect to relational *epistemic* properties. Thus, unlike the proposals briefly discussed in this section, the proposal offered above suggests that the difference between reduced and reducing item stems from non-cognitive, non-epistemic, and non-pragmatic semantic differences between the reduced and the reducing item. The 2-D characterization hinted at in the previous section grounds the reductive relation in epistemological properties of the reduced and the reducing item as well; here, conceptual or semantic aspects are *described in epistemological terminology*. The next section discusses one problem these views face.

5.7.3 Reductive Dependence as Grounding Epistemic and Pragmatic Differences

How a theory of Fregean sense should look like is a question that has to be settled on independent grounds. Thus, what ultimately is the correct version of the definition of reduction will, hopefully, fall out of our correct semantic theory. Just like the question of what theories basically are will decide, to some extent, which version of a characterization of theory-reduction is more appropriate – a syntactic or a semantic one – a theory of what Fregean senses basically are will ultimately decide how to account for the features mimicked by the notion of a property structure. Thus, as long as they preserve the structure of the characterization given above, different accounts may look equally appropriate, as long as we ignore our semantic prejudices and biases operating in the background of our philosophical judgments. Let me articulate one of my fundamental semantic convictions, and then argue that this conviction nicely matches an implicit commitment in the reduction debate and, thus, is perfectly suited to shape an appropriate explication of reduction, which is supposed to reveal such implicit commitments.

The concept of H_2O gives a better epistemic access to water (at least in some respects) than the concept of water (if there is such a concept). Grasping the concept of H_2O, and knowing that a given sample of water is H_2O may come with a practical advantage – we improve our chances to successfully manipulate water (see Sect. 9.4). But does this give reason to assume that the conceptual difference is basically epistemological, cognitive or pragmatic *in nature*? Conceptual differences are not to be conflated with epistemic or pragmatic differences. The difference between water and H_2O is conceptual in nature. Conceptual differences *explain* the relevant epistemic and pragmatic differences. On this view, there is an epistemic and a pragmatic difference *because* there is a conceptual difference. This is what I take to be part of the Fregean view that difference in sense *explains*, and is not to be conflated with difference in cognitive role (Frege 1892, for a discussion, see (van Riel 2011)); it is opposed to semantic theories that tie notions such as Fregean sense or conceptual content to epistemic notions (such as Chalmers 2002, 2004, 2006). Accordingly, once we have fixed the relevant conceptual difference, we should be in a position to illuminate the resulting epistemic difference. This is why this conviction is particularly well suited to give a characterization of reduction, in the sense that it captures an assumption at least implicit in the reduction debate.

It has been widely assumed that part of the benefits of reductions is that they go together with scientific *progress*. This progress may very well be epistemic or pragmatic in nature. Moving from folk chemistry to chemistry that employs a formal apparatus is good because, as Van Gulick suggests, the representational framwork is cognitively more appropriate and has the capacity to lead to pragmatic success. But is this a brute fact? What is responsible for this difference in epistemic, cognitive and pragmatic success? On pragmatic and epistemic accounts, the notion of reduction won't tell us. It will just tell us *that* there is an epistemic or pragmatic difference. It will not tell us what this difference depends on. So, if the commitment to the idea that in principle, reduction goes together with epistemic and pragmatic progress, and if such progress is not a brute fact but rather depends on properties of the reduced and the reducing item, then reference to semantic features, which are not, in turn, explicated in pragmatic or epistemic terms, forms a good starting point. To this extent, the proposal offered above is more powerful than its rivals. Epistemic, cognitive and pragmatic differences between knowledge that a given sample of water is water, and knowledge that a given sample of water is H_2O result from, and are not to be conflated with, semantic differences between 'water' and 'H_2O'. The different cognitive, pragmatic, or a priori roles of these expressions are grounded in, and are not to be conflated with, their respective conceptual contents. This assumption may very well turn out to be false; but this has to be decided on independent grounds, and the explication proposed here could easily be adjusted.

5.8 Conclusion

We have defined an appropriate core notion of reduction, which captures the characterizations offered before, and three important derivative notions. The core notion matches the job description, and it is motivated by the solution to the puzzle offered in Chap. 3 as well as by the alleged similarity between mechanistic and reductive explanation.

Q1: How can we reconcile *diversity* and *directionality* with strong *unity*?

...

Th. 4: A core notion and derivative notions of reduction can be explicated; the explication fits paradigmatic cases, is coherent (solves the puzzle) and satisfies an intuitive job-description.

...

The explication proposed here uses a model of property structures. Alternative ways of capturing a similar idea are easily conceivable. However, it has been argued that the version proposed here is (i) modest, in that property structures are said to be determined by whatever plays the role of Fregean senses, and (ii) it offers the resources to explain epistemic, cognitive and pragmatic differences between different conceptual contents – a feature I take to be an advantage within the context of the reduction debate.

We can now move on to further motivate the proposal, applying it to the reduction debates in the philosophy of mind and the philosophy of science, as well as related issues in Part II.[20] The following Appendix contains a semi-formal treatment of property structure terms (it will not be referred to in the remainder of the book).

[20]Note that the distinction between two reduction-debates (one in the philosophy of science, another in the philosophy of mind) is based on an idealization. The discussion in the philosophy of mind was strongly influenced by early models of reduction developed in the philosophy of science. However, we can distinguish between two different *foci* or *tendencies* in the two fields: In the post-Nagelian debate on reduction in the philosophy of science, the discussion was often though not invariably driven by *extension first* approaches. This is suggested not only by the examples which are given and which are intensively discussed, but also by the explicitly mentioned goal to describe *actual* theory-succession (Kemeny and Oppenheim 1956; Wimsatt 1974; Sneed 1971). This is why in the post-Nagelian debate on reduction in the philosophy of science, *replacement* played such a crucial role, and identity-based reduction did not receive special attention. Contrary to that, in the philosophy of mind the notion of reduction attracted attention partly because it seemed to give a *metaphysical* picture of the relation between the mind and underlying processes – a relation completely independent of actual scientific developments. This is, again, pertinent in some branches of the Philosophy of Science – namely, when issues regarding unity are discussed (Carnap 1934; Oppenheim and Putnam 1958; Causey 1977; Cartwright 1999). This motivates the idea of drawing the distinction between one reduction-debate in the philosophy of science and another in the philosophy of mind.

Appendix: A Semi-formal Treatment of Property Structure Terms

This appendix introduces a semi-formal treatment of property structure terms (‘*PST*s’). Only a limited number of cases of PSTs will be covered by this semi-formal treatment. However, the overall strategy should become apparent. The semi-formal treatment of these terms will be introduced as follows: In a first step, a *schema structure* (S_{ps}) for property structure terms will be introduced. A schema structure contains a list of schematic expressions and syntactic rules to form complex schematic expressions, and, here, it contains additional operators that enable us to treat natural language expressions in a relatively uniform way. In this, it is similar to the characterization of the syntax of a formal language. However, the expressions obtainable from a schema-structure will not be given an interpretation; rather, their instances *come* with a meaning – instances of schemata obtainable from this schema-structure are natural language expressions or expressions generated by the application of operators to natural language expressions. The semantic values of the results of the application of these operators crucially depend upon the meanings of the natural language expressions these operators are applied to, and they will be given in a meta-linguistic fashion, referring to the meaning of the expression they are applied to.

A Set of Rules for Obtaining Schemata for Property Structure Terms

The schema structure will be called ‘S_{ps}’. It contains schema letters and additional operators.

Schema-letters of S_{ps} and operators:

- Variables ‘x’, ‘y’, with or without index (with the substitution class {‘x^1’, ‘x^2’, ... ‘x^n’, ‘y^1’, ‘y^2’ ... ‘y^n’}.
- Predicate letters ‘P^n’, ‘Q^n’ (with $n \geq 1$, with or without index) for natural language predicates that are simple in the sense that they do not contain predicates or singular terms as constituents.
- Letters for kind terms ‘A’, with or without index, which do not contain predicates or other terms as constituents.

Additional Operators

- ‘$< \ldots, \ldots >$’ (which represents the identically looking sign for n-tuples),
- ‘[. . .]’ (stands for ‘[. . .]’ – an operator that works on predicates the argument positions of which are taken by free variables; it is similar to operators ‘the property signified by ‘ . . . ’’ and similar operators),

- ‘&’ (designating the function on properties and states ‘and’ determines)[21]
- EXISTS (designating the function on properties and states which is determined by the existential quantifier)
- The set of operators that are expressed by expressions of the form ‘... $^{\xi 1}$’ (Putting a variable in this place after an expression, we express a function from properties to properties or states of affairs, a function that does not have a natural language counterpart; in natural language, this function is captured by arrangement: the property signified by ‘... loves Peter’ is distinct from the property signified by ‘Peter loves ...’. In our schema, we need to distinguish between these properties. Instead of relying on arrangement, we introduce this set of operators. They mimic the syntactic function of arrangement of expressions occupying argument-positions within complex predicates.)

Note that this fragment does not contain place-holders for modal operators, or something that corresponds to an operator like ‘the fact that ...’ (which would make things unnecessary complex) and a considerable number of other operators which might be relevant for scientific discourse. Thus, we should regard the account presented here as a partial account of property structures. We can now recursively define the set of property-structure *term*-schemata. I shall talk about ‘terms’ from now on *simpliciter*. It should be noted, however, that these terms are *schemata* for terms.

Terms (term-schemata) in S_{ps}:

- If Φ^n is an n-place predicate (with $n \geq 1$) and $\xi_1 \ldots \xi_n$ are variables, then ‘$[\Phi\xi_1 \ldots \xi_n]$’ is a *term*.

The goal of this rule is to build an operator that takes a predicate as an argument. The result is a term referring to the property signified by the predicate.

- If α is a kind term, then α is a *term*.

Kind terms just contribute the kind they designate to property structures. The following two rules give a procedure to model conjunction and the existential quantifier.

- If α_1 and α_2 are *terms*, then $<\&, <\alpha_1, \alpha_2>>$ is a *term*.
- If α is a *term* and ξ_1 is a variable which is free in α, then $<\mathrm{EXIST}\ \xi_1, \alpha>$ is a *term*.

Here, the variable ξ is bound by a quantifier EXIST in an instance of a schema of the form $<\mathrm{EXIST}\ \xi, <\alpha>>$ iff it occurs within the scope of the quantifier, namely, within ‘$<\alpha>$’.

- If α is a *term* with free variables ξ_1-ξ_n, and $\beta_1 \ldots \beta_m$ are *terms* and $n \geq m$, then $<\alpha, <\beta_1\xi^1, \ldots \beta_m\xi^m>>$ is a *term*.

---

[21]Other connectives are ignored here to keep things simple. However, other connectives can be introduced in a similar fashion.

This is supposed to capture the idea that we can generate complex expressions, where 'a' is obtained from an n-ary predicate with free argument positions $\xi_1 - \xi_n$, some or all of which are taken by '$\beta_1\xi^1$', ... '$\beta_m\xi^m$'. The resulting property structure term will model the property structure of a predicate whose *arity* is $n - m$ or, in case that $n - m = 0$, of a sentence. The final clause states that all and only those terms obtainable from this procedure are property structure terms.

- All and only those *terms* obtainable in a finite number of steps from the previous rules are *terms* in S_{ps}.

Now, for example, the expression '<&, <[x is a ball], [x is red]>>' is an instance of a schema obtainable from this schema structure, namely: <&, <[Px], [Qx]>>. Similarly '<EXIST x, <[x is an instance of y], <[z is wise]$^{y>}$>>' is an instance of: <EXIST x, <[Px,y], <α^{y}>>>.

What about the semantics of instances of schemata obtainable from the schema structure just given?

Semantic characterizations of Instances of S_{ps} Schemata

For simple predicates and simple singular terms, it is not complicated to give the semantic values – they come with the meaning they have, and if they designate or signify something, then their contribution to the ontological structure can intuitively be made precise as follows: The operator '[...]' takes predicates and refers to the property the predicate signifies, if any. Kind terms contribute the kind they designate. Thus, the property structure we get access to by an instance of (an instance of) '[Φ $\xi_1 \ldots \xi_n$]' is the property signified by the predicate that is substituted for the instance of Φ. But what about quantifiers or what is treated in predicate logic as sentential operators? The basic idea is to treat them as designating partial functions that map properties (or other objects) onto properties (and other objects). The general idea for this procedure has been spelled out above (Chap. 5). What has been taken as being merely *determined* by the relevant natural language expression will here be treated as the semantic value of the expressions generated by the application of the additional operators.

We now characterize an interpretation function I_{ps}. This function specifies the semantics for simple expressions that conform to the schema structure given above. Let D^P be the set of properties, D^F the set of states of affairs, D^I the set of individuals, and $D^{PF} = D^P \cup D^F$ and $D = D^P \cup D^F \cup D^I$, and D^{Pred} be the set of simple predicates. We then get the following characterization of our interpretation function (let 'α' stand for kind terms).

(Characterization Interpretation Function I_{ps})	– $I_{ps}(\alpha)$ = *the entity designated by* α *(if any).* – $I_{ps}([\ldots]) = f: D^{Pred} \rightarrow D^{P}$, *such that f(x) = y* $\in$ *D or f(x)is undefined. (f(x) = y, such that x signifies y, if anything).*[22] – $I_{ps}(\&) = f: D^{PF} \times D^{PF} \rightarrow D^{PF}$, *such that f(x) = y* $\in D^{PF}$ *or f(x)is undefined. (f(x) = the significatum of the appropriate instance of '*y_1 *and* y_2*', with x = <*z_1, z_2*> and the instance of* y_1 *signifies or designates* z_1 *and the instance of* y_2 *designates or signifies* z_2*, if anything).* – $I_{ps}(EXIST) = f: D^{P} \rightarrow D^{PF}$, *such that f(x) = y* $\in D^{PF}$ *or f(x) is undefined. (f(x) = the significatum of the appropriate instance of 'There is at least some y such that Fy' with F signifying x, if anything).* – $I_{ps}(\xi^1 \ldots \xi^n) = f: D^{P} \rightarrow D^{PF}$, *such that f(x) = y* $\in D^{PF}$ *or f(x) is undefined. (f(x) = the significatum of any appropriate instance of 'Fa,* y_n*, ...* y_m*' (n/m* $\geq$ *0), with x = the property signified by 'Fa', if anything)*

The tuple-operator, '< . . . >', works as usual.

We can now define the notion of a property structure:

(Def. property structure of an expression):	*A property structure an expression E gives access to is the referent of an instance of a schema generated in* S_{ps} *obtained from E.*

Note again that S_{ps} is limited; thus, it should be extended whenever necessary. Again, property structures and their complex and simple constituents sometimes determine properties or states in a specific way; namely, they determine what is signified, referred to or stated by the expression they are obtained from (if this expression signifies, refers to or states anything), given that this is fully determined by the property structure in the following way (to mark the difference to the characterization above, an asterisk is added):

(Def. Determination):* *A property structure PS determines an entity x as follows:*

- *Iff PS takes the form a and 'a' is a kind term, then a = x.*
- *Iff PS takes the form [Φ . . .], then [Φ . . .] = x.*
- *Iff PS takes the form <f<*a_1*, ...* a_n*>> and* a_1 *determines* x_1 *... and* a_n *determines* x_n *then f(<*x_1 *...* x_n*>) = x.*

[22]Thus, we assume that any property that could enter a property structure is expressible.

- *Iff PS takes the form* $<a, <\beta_1\xi^1, \ldots \beta_n\xi^n>$ *such that* a *determines the m-ary property* y *and* β_1 *determines* $x_1 \ldots$ *and* β_n *determines* x_n *then*
  - *if* $m - n > 0$, *then* $x =$ *the* $(m - n)$*-ary property of standing in* y *to* $x_1, \ldots, x_n$*,, or*
  - *if* $m - n = 0$, *then* $x =$ *the fact that* $x_1 \ldots x_n$ *instantiate* y.

Note that we lack an operator similar to the iota-operator. Intuitively, introduction of such an expression would enable us to generate property structures that determine individuals. The notion of satisfaction can be introduced in a way similar to the way it was given above: the extension, or the extension's members satisfy the property structure. Now, let us turn to an example in order to make these ideas more precise, and to illustrate how these tuples relate to reductive explanation.

An Example

Consider, again, the following two expressions.

'Water'; 'H_2O'.

Needless to say: It will be taken for granted that H_2O = water. Let 'water' be a kind term which does not have a meaning, but only a referent. Then we get the following property structure the expression gives access to in its ordinary use:

[x is water].

Now, assume that 'water' has a meaning, given by the predicate '_is tasteless and liquid'. We would get:

< [&], <[x is tasteless], [x is liquid]>>

'H_2O', in contrast, gives access to a property structure that reveals its chemical constitution. Intuitively, it is the property which is instantiated by entities that have the property of having a proper part that is an oxygen atom and two proper parts that are hydrogen atoms.[23] We can then compare these distinct PSs (of 'water' and 'H_2O') with respect to their basic properties:

---

[23]To model the property structure 'H_2O' gives access to, we should add the connective '→' (if, then) and the general quantifier 'EVERY' to be characterized in a way similar to 'and' and 'EXIST'. Let '_ is a PP of _' be '_is a proper part of_'.

<EVERY y, <→, <[x is instantiated by y], <EXIST z_1,z_2, z_3, <&, <&, <[z_1 is a PP of y], <&, <[z_2 is a PP of y], [z_3 is a PP of y]>>>>, <&, <[z_1 is hydrogen], <&, <[z_2 is hydrogen], [z_3 is oxygen]>>>>>>>>

This version still neglects some details, because the language still lacks expressive power. We should add:

<¬, <EXIST a z* <[z* is a PP of y], ¬ [z* is identical to z_1], ¬[z* is identical to z_2], ¬[z* is identical to z_3]>.

On the one hand we have *being water* or *being tasteless* and *being liquid*. The PS of 'H_2O' will contain:

[x is hydrogen], [x is oxygen], [x is a proper part of y].

These are clearly properties of constituents of water-molecules. Moreover, these are properties that are explanatorily relevant for why water behaves the way it does, and, according to some interpretations, that explain to a certain extent what water basically is. This parallels the idea underlying mechanistic explanation: We explain a whole in terms of the causally relevant organization of its parts. Thus, we can explain why some *explanans* is appropriate for an *explanandum* referring to the property structure under which it gives access to the target of the *explanandum*: In virtue of giving us a constitutive structure, the *explanans* is appropriate.

Thus, we have outlined a semi-formal treatment of property-structure terms that gives a more precise idea of the metaphysics of property structures. Let me finish these remarks mentioning that the notion of a tuple is nevertheless problematic in this context: There can be many property structure terms that are instances of a schema obtainable from S_{ps} for one non-ambiguous natural language expression. For example, a conjunction with only two conjuncts can be modeled in two different ways: a natural language predicate of the form '_is F &_ is G' can be modeled as '<&<[Fy], [Gx]>>' and as '<&<[Gy], [Fx]>>'. Now, intuitively, property structures are less fine grained than tuples. Thus, these tuples should function as a mere model for property structures only.

Bibliography

Bealer, George. 1993. A solution to Frege's puzzle. *Philosophical Perspectives* 7: 17–60.

Bechtel, William. 2007. Reducing psychology while maintaining its autonomy via mechanistic explanation. In *The matter of the mind: Philosophical essays on psychology, neuroscience and reduction*, ed. Maurice Schouten and Huib Looren de Jong, 172–198. Oxford: Basil Blackwell.

Black, Max. 1944. Paradox of analysis. *Mind* 53: 263–267.

Carnap, Rudolf. 1934. *The unity of science*. London: Kegan Paul, Trench, Trubner, & Co.

This ensures that within the limits of our schema structure, we fix that water-instances do not have more than three constituents. Moreover, we should add <&, <<¬, [$z_1$is identical to z_2]>,<¬, [$z_2$is identical to z_3]>>> to ensure that H_2O has three constituents. Alternatively, we could introduce a further operator, namely, 'there are exactly three'. Note that being water is the property that is determined by this structure: The leftmost operator ('EVERY y') contributes a function that maps the property signified by 'if _is instantiated by _ then there is a z_1, z_2, z_3 such that z_1 is a proper part of _ and z_2 is a proper part of _ and z_3 is a proper part of _and z_1 is hydrogen and z_2 is hydrogen and z_3 is oxygen' onto the property signified by 'for any y, if y instantiates _, then there is a z_1, z_2, z_3 such that z_1 is a proper part of y and z_2 is a proper part y of and z_3 is a proper part of y and z_1 is hydrogen and z_2 is hydrogen and z_3 is oxygen'. This predicate, however, signifies a property which is instantiated by the property of being water *only*. It is satisfied by what is water – the extension of the property.

Cartwright, Nancy. 1999. *The dappled world: A study of the boundaries of science*. Cambridge: Cambridge University Press.
Causey, Robert. 1977. *Unity of science*. Dordrecht: Reidel.
Chalmers, David. 1996. *The conscious mind*. Oxford: Oxford University Press.
Chalmers, David. 2002. On sense and intension. *Philosophical Perspectives* 16: 135–182.
Chalmers, David. 2004. Epistemic two-dimensional semantics. *Philosophical Studies* 118: 153–226.
Chalmers, David. 2006. The foundations of two-dimensional semantics. In *Two-dimensional semantics: Foundations and applications*, ed. José Macia and Manuel Garcia-Carpintero, 55–140. Oxford: Oxford University Press.
Chalmers, David. 2011. Propositions and attitude ascriptions. A Fregean account. *Noûs* 45: 595–639.
Church, Alonzo. 1973. Outline of a revised formulation of the logic of sense and denotation (Part I). *Noûs* 7: 24–33.
Church, Alonzo. 1974. Outline of a revised formulation of the logic of sense and denotation (Part II). *Noûs* 8: 135–156.
Church, Alonzo. 1993. A revised formulation of the logic of sense and denotation. Alternative (1). *Noûs* 27: 141–157.
Correia, Fabrice. 2008. Ontological dependence. *Philosophy Compass* 3(5): 1013–1032.
Craver, Carl. 2005. Beyond reduction: Mechanisms, multifield integration and the unity of neuroscience. *Studies in the History and Philosophy of Biological and Biomedical Sciences* 36: 373–395.
Craver, Carl. 2007. *Explaining the brain: Mechanisms and the mosaic unity of neuroscience*. Oxford: Oxford University Press/Clarendon Press.
Cresswell, John. 1985. *Structured meanings. The semantics of propositional attitudes*. Cambridge, MA: MIT Press.
Darden, Lindley, and Nancy Maull. 1977. Interfield theories. *Philosophy of Science* 44: 43–64.
Enç, Berent. 1976. Identity statements and microreductions. *Journal of Philosophy* 73: 285–306.
Fodor, Jerry A. 1974. Special sciences: Or the disunity of science as a working hypothesis. *Synthese* 28: 97–115.
Frege, Gottlob. 1892. Über Sinn und Bedeutung. *Zeitschrift für Philosophie und philosophische Kritik* 100: 25–50.
Horst, Steven. 2007. *Beyond reduction: Philosophy of mind and post-reductionist philosophy of science*. Oxford: Oxford University Press.
Kemeny, John G., and Paul Oppenheim. 1956. On reduction. *Philosophical Studies* 7: 6–19.
Kim, Jaegwon. 2006. *Philosophy of mind*. Boulder: Westview Press.
Kim, Jaegwon. 2008. Reduction and reductive explanation. Is one possible without the other? In *Being reduced. New essays on reduction, explanation and causation*, ed. Jesper Kallestrup and Jakob Hohwy, 93–114. Oxford: Oxford University Press.
Kripke, Saul. 1980. *Naming and necessity*. Cambridge, MA: Harvard University Press.
Künne, Wolfgang. 1983. *Abstrakte Gegenstände. Semantik und Ontologie*. Frankfurt: Suhrkamp.
Künne, Wolfgang. 1990. Perception, fiction, and elliptical speech. In *Thinking and the structure of the world. Hector-Neri Castañeda's epistemic ontology presented and criticised*, ed. Pape Helmut and Jacobi Klaus, 259–267. Berlin: de Gruyter.
Künne, Wofgang. 2001. Constituents of concepts. Bolzano vs. Frege. In *Building on Frege. New essays on sense, content, and concept*, ed. Albert Newen, Ulrich Nortmann, and Rainer Stuhlmann-Laeisz, 267–286. Stanford: CSLI Publishing.
Nagel, Ernest. 1961. *The structure of science. Problems in the logic of explanation*. New York: Harcourt, Brace & World, Inc.
Newen, Albert. 1996. *Kontext, Referenz und Bedeutung. Eine Bedeutungstheorie singulärer Terme*. Paderborn: Schöningh-Verlag.
Oppenheim, Paul, and Hilary Putnam. 1958. The unity of science as a working hypothesis. In *Concepts, theories, and the mind-body problem*, ed. Grover Maxwell, Herbert Feigl, and Michael Scriven, 3–36. Minneapolis: Minnesota University Press.

Quine, Willard V.O. 1951. Two dogmas of empiricism. *Philosophical Review* 60: 20–43.
Quine, Willard V.O. 1980. The variable and its place in reference. In *Philosophical subjects – Essays presented to P. F. Strawson*, ed. Zak van Straaten, 164–173. Oxford: Clarendon.
Sarkar, Sahotra. 1992. Models of reduction and categories of reductionism. *Synthese* 91: 167–194.
Schaffner, Kenneth. 1993. *Discovery and explanation in biology and medicine*. Chicago: The University of Chicago Press.
Schaffner, Kenneth. 2006. Reduction: The Cheshire Cat problem and a return to roots. *Synthese* 151: 377–402.
Schaffner, Kenneth. 2012. Ernes Nagel and reduction. *Journal of Philosophy* 109: 534–565.
Schiffer, Stephen. 2003. *The things we mean*. Oxford: Oxford University Press.
Schnieder, Benjamin. 2006. Attributing properties. *American Philosophical Quarterly* 43: 315–328.
Smart, John J.C. 1959. Sensations and brain processes. *Philosophical Review* 68: 141–156.
Sneed, Joseph. 1971. *The logical structure of mathematical physics*. Dordrecht: Reidel.
Strawson, Peter F. 1974. *Subject and predicate in logic and grammar*. London: Methuen.
Tichy, Pavel. 1988. *The foundations of Frege's logic*. Berlin: Walter de Gruyter.
van Fraassen, Bas. 1980. *The scientific image*. Oxford: Clarendon.
Van Gulick, Robert N. 1992. Nonreductive materialism and the nature of intertheoretical constraint. In *Emergence or reduction? essays on the prospects of non-reductive physicalism*, ed. Ansgar Beckermann, Hans Flohr, and Jaegwon Kim, 157–179. Berlin: de Gruyter.
Van Gulick, Robert N. 2001. Reduction, emergence and other recent options on the mind/body problem: A philosophic overview. *Journal of Consciousness Studies* 8: 1–34.
van Riel, Raphael. 2011. Cognitive significance and epistemic intensions. *Logique et Analyse* 54: 503–516.
Walton, Kendal L. 1990. *Mimesis as make-believe: On the foundations of the representational arts*. Cambridge, MA: Harvard University Press.
Wimsatt, William. 1974. Complexity and organization. In *PSA: Proceedings of the biennial meeting of the Philosophy of Science Association,* 1972, Boston studies in the philosophy of science, ed. Kenneth Schaffner and R. Cohen, 67–86. Dordrecht: Reidel.
Wimsatt, William C. 2006. Aggregate, composed, and evolved systems: Reductionistic heuristics as means to more holistic theories. *Biology and Philosophy* 21: 667–702.
Woodward, James. 2003. *Making things happen. A theory of causal explanation*. Oxford: Oxford University Press.

Part II
The Explication at Work

Chapter 6
Reduction in the Philosophy of Mind

Now it is time to show the relevance of the notion and the fruitfulness of the explication. The concept of reduction defined in the previous chapter is key to an understanding of the reduction-debate in the philosophy of mind. Reductive identity-theories, type-identity theories as well as token-identity theories, and models of functional reduction conceive of reduction precisely as described above: These theories are supposed to reconcile strong unity with diversity and directionality; according to this conception, a reduces to b only if $a = b$.

6.1 Background

The connection between identity theses of the mind and reduction is quite obvious: Issues of reduction were often discussed *as* issues of identification. Early identity-theorists like Smart (1959, 1963), Feigl (1967) and Place (1956, 1960) focused on identification of mental types with lower level ones, even though their target clearly was that of (in-principle) reducibility (and, thus, reductionism). Similarly, those who resisted the idea of type-identity (most famously: Putnam 1967; Fodor 1974, 1997), argued against identification of the mental with some other level; again, the target clearly was reducibility. These discussions usually took and take place within a functionalist framework.

The present chapter proceeds in two parts. In Sects. 6.2, 6.3 and 6.4, I will walk us through the successive stages of different versions of identity-theories about the mind, in order to shed light onto their relevance regarding issues of identity-based reduction, and to show how the explication proposed here can be applied to these versions of identity-theories. Section 6.5 summarizes the main points, before turning to a characterization of functional reduction in Sect. 6.6, where functional reduction will be tied to identity-based reduction.

As we shall see below (Chaps. 7 and 8), reduction in the philosophy of science is usually conceived of as a relation that holds primarily between theories. Not so in the philosophy of mind. In the philosophy of mind, the stance towards possible *relata* of

R. van Riel, *The Concept of Reduction*, Philosophical Studies Series 121, DOI 10.1007/978-3-319-04162-9_6, © Springer International Publishing Switzerland 2014 the reduction relation is influenced by ontologically liberal philosophical traditions, which allow for talking freely about *properties*, *phenomena*, *states of affairs* and so forth; accordingly, all these items figure in the list of candidates for the *relata* of the reduction relation. In this sense, the application of the explication proposed above is far more straightforward in the context of the philosophy of mind than in the philosophy of science. Note that in what follows, we will be concerned with reductionism concerning the *mental*. It should be noted, however, that any other form of reductionism, say, about biology and biological kinds, could come in a similar format.

6.2 Type-Identity Theory

Type-identity theory is reductive. It is not merely the theory that the mental is subsumed under the physical or the physiological. The latter is assumed to be more *fundamental*, or *prior* to the former. Put differently: Subscribing to type-identity theory, in this sense, you oppose idealism as well as a neutral stance towards what is more fundamental. To get an idea of different versions of type-identity theory, let us reflect upon a quote taken form Herbert Feigl's (1967). Feigl gives the following characterization of type-identity theory as some version of reductionism (Feigl 1967, 71 ff.) answering the question of "whether the mental and the physical can in some sense be identified":

> [I]t is proper to speak of "*identification* [. . .]" Concepts of molar behavior theory like habit strength, expectancy, drive, instinct, memory trace, repression, superego, etc., may yet be identified in a future psychophysiology with specific types of neural-structure-and-process-patterns. (Feigl 1967, 77)

Reference to 'a future psychophysiology' makes this interpretation of type-identity-theory an epistemological kind of reductionism, in at least one respect: Reduction is interpreted as being possible not only in some principled way, but rather as being possibly carried out by *us*, as being possible given our epistemological capacities. Talking about *concepts*, Feigl seems to have properties in mind, properties that are to be identified with the relevant *scientific kinds*. Combining the claim of identity with the list of examples given in this citation, we can derive the following thesis: *Mental kinds are identical with physical kinds*, which is offered as an explanation of the claim that *the mental and the physical can be identified*, or, perhaps less misleadingly: *The mental is to be subsumed under the physical*. This thesis is true only if the relevant individual identity statements are true. However, it is not clear whether or not type-identity theory is fully captured by the relevant identity-statements. Most formulations of type identity-theory are intuitive, and they allow for different interpretations. They all have in common that they rely on a strong interpretation of unity – according to type-*identity* theories, there is unity among mental and physiological kinds in the sense that every mental kind is identical to a physiological kind. In a first approximation, the type-identity thesis can be stated as follows: *For every x, if x is a mental kind, then there is a y such that y is physical (physiological, neural ...) and x = y.* This thesis, however, is a bad candidate for type-identity theory, for it does not rule out the possibility that type-identity theory is contingently true. According to this formulation (in conjunction with a denial of stronger claims), any actual mental type is identical to a physical type (which is metaphysically necessary), but there could have been (it is even nomologically possible that there is) at least one mental type that is not identical to any physical type. Why is this interpretation wrong-headed?

Type-identity theorists seem to justify their claims referring to a specific sort of inductive reasoning: Scientific findings justify a claim about *the nomological structure* of our world – nicely formulated by Smart (1959, 142f.) who explicitly appeals to *Ockham's razor*: In the light of what science tells us, the laws that would govern the non-physical, mental sphere must look rather odd, if type identity theory were false. That is, this sort of materialism is based on an assumption about what we should expect our world to be shaped according to science, and, thus, *according to the laws*. But the shape of our world, or its nomological or metaphysical structure is not to be conflated with what actually though contingently exists or is instantiated (this would best be represented by a *list* of objects). Thus, we should conclude that the non-modalized thesis is not a serious candidate for an articulation of type-identity theory. It is, in a sense, an interesting ontological claim about the set of actually existing types. However, it does not fully capture the thesis that is justified by the line of reasoning proposed by the type-identity theorist.

So, what about alternative interpretations of the modalities involved here? According to one reading, we would get a thesis denying the metaphysical possibility of immaterial minds or souls. This is the case if we talk about possibly mental kinds *whatsoever*. According to yet another alternative, we get a claim about all nomologically possible mental types, and I think that this is what the reductionist needs: At least, any mental kind that is possibly instantiated according to the laws of nature is identical to a physical kind. The strong claim that immaterial or non-physical minds are metaphysically impossible is not what we should focus on – what happens in remote and, arguably, queer possible worlds is not our concern here. The question of reduction, as I understand the claim of type-identity theory and as it seems connected to the endeavor of science, is concerned with exactly those mental kinds that are nomologically possibly instantiated. So, here is our candidate:

(Type-Identity-Thesis): Any mental kind which is nomologically possibly instantiated is identical to a physical kind.[1]

[1]Note that a metaphysical thesis, a thesis including nomologically possible or metaphysically possible types might turn out to be equivalent to this thesis iff the set of actual types is identical to the set of nomologically possible types and is identical to the set of metaphysically possible types. However, even in this case, the line of reasoning in favor of type-identity theory would be as follows: In virtue of the fact that all nomologically possible types are physical types, any actual type is physical, and in virtue of the fact that all nomologically possible types are physical and they form the set of metaphysically possible types, any metaphysically possible type is physical.

This is a thesis about *mental* kinds. Reduction-talk is not conceptually tied to the mental: Kinds other than mental may reduce as well. If type-identity theory is relevant for reduction then we need an extension of type-identity theory which is more liberal concerning the relevant set of kinds: It should at least be about any nomologically possible natural kind whatsoever. For some other partial reductionism, we can define a similar claim, concerning kinds of a specific sort, for example, for biological kinds. As such, these characterizations do not give us an idea of how to conceive of directionality or diversity. Type identity theory, thus construed, is underdetermined – it just does not express a *reductive* thesis in any interesting sense. At best, it gives a partial explication of a reductive thesis. Building on the explication proposed above, we can easily fix this. It allows us to define a derivative notion of type-reduction, by explicitly requiring that the objects designated by expressions that flank the reduction predicate in true reduction statements be types:

(Explication VII – Type-Reduction) *x, when presented under PS1,* ***type-reduces*** *to x, when presented under PS2 iff*

(i) *x is a type &*
(ii) *PS2 is a constitutive structure of x with respect to PS1.*

This gives an idea of how to account for reduction statements about individual types. The thesis that mental types reduces to physical types is best construed in terms of generic statements, and, hence, in terms of generic reduction. This explication accounts for diversity in terms of *conceptual* diversity. Sections 6.3 and 6.5 argue that this perfectly matches and vindicates a number of points that have been made in the debate on type-identity theory. The next two sections discuss the connection between conceptual and metaphysical issues in type identity theories, relating them to the explication proposed above.

6.3 Type-Identity Theory and Conceptual Diversity

In the debate on type-identity theory, the topic of conceptual diversity has been touched on several occasions. Smart (1959) discussed a number of possible objections to type-identity theory, which in fact merely attack a strong version of type-identity theory, rather than the version he, and, according to him, Place tried to defend. These alleged counter-arguments roughly run as follows (Smart 1959, 146 ff.): There is no conceptual route from the mental to the physical, and, therefore, the relevant identity-claims are mistaken. If $a = b$, then it is conceptually necessary

---

Thus, even here, the nomological thesis is prior to its equivalent (though not identical) counterparts. Something similar holds for the weaker claim that the set of actual types is the set of nomologically possible types.

that $a = b$. This is plainly false – there can be informative identity claims, and Frege (1892) gave a model of how this is possible. The basic idea is that every expression can be used to perform two tasks: picking out or designating an object and expressing a conceptual content, or, in Fregean terms, a *sense* or a *mode of presentation*. The latter is responsible for cognitive role. An identity claim can then be informative and true iff the conceptual contents of the two expressions flanking the identity-sign differ appropriately (which makes it informative) whilst at the same time the designatum of the first just is the designatum of the second (which makes it true). So, the conditional – *if* $a = b$ then it is conceptually necessary that $a = b$ – is false. Smart (1959) explicitly based his reply to the objections he considered on this Fregean idea. Given the criticism Smart raised against the assumption that the relevant identity-claims are a priori, it is worth noting that the relevant identity-statements are supposed to be *informative*. This falls out of the explication given above, because it relies on the assumption that the relevant diversity is conceptual in nature: Insofar as the a priori roles of the concept of water and the concept of H_2O are mimicked by property structures, and given that epistemic possibility does not track metaphysical possibility, it is obvious that the corresponding identity statement, that water = H_2O, is informative.

The way Smart's anonymous opponent conceives of the connection between identity statements and conceptual necessity is, to some extent, mimicked by a way of talking according to which kinds or types are more kin to concepts as conceived of here than to non-representational entities. The idea that there is a conceptual connection between the mental and the physical is tied to a *façon de parler* that can be found in various parts of the debate on reductionism and type-identity theory, according to which we can *analyze* kinds, events or states in functional terms (*cf.* Block 1995; Kim 2005, 167; Jackson 2005). Type-identity theorists hold that mental types or kinds or properties are identical to physical types or kinds or properties. This suggests that *mental kinds* and *physical kinds* are non-representational objects – an interpretation that seems rather plausible given that type-identity theory is a metaphysical theory. Necessarily, there are no conceptual connections between kinds (or non-representational worldly entities), just because kinds are not the right sort of entities to instantiate conceptual relations in any non-derivative way.[2] Kinds are the subject of ontological claims about mental properties,

[2]Also in the philosophy of science, the difference between concepts as representational devices and 'concept' as a stylistic variant of 'property' make it sometimes hard to uncover what philosophers have in mind when they employ the term. In his (1990) Paul Thagard introduced a model of theory-change in the sense of theory replacement. Discussing the example of the revolution associated with Lavoisier's oxygen-theory, Thagard gives a detailed analysis of what he labels 'conceptual' change. Unfortunately, it is rather difficult to uncover what a *concept* (the constituent of a conceptual scheme) is supposed to be in this context. The only hints we get stem from his descriptions of the relevant relations between conceptual items: They are described in terms of *being an instance of*, as a *part-whole relation*, and as a *relation between properties and kinds* (Thagard 1990, 184 f.). Thus, it seems that Thagard implicitly adopts an ontological theory of concepts – a theory according to which the concept is individuated by the property, or kind, or

about the causal role of the mental and the like. Thus, they are not to be conflated with representational entities – they belong to the (concrete or abstract) worldly stuff (first order) representational entities like (first order) concepts or linguistic patterns are *about*. The same kind can be represented by two different conceptual contents. The conceptual contents of 'the kind cat' and of 'the kind of animals the extension of which forms the set of my childrens' favorite animals' pick out the same kind, but they are associated with completely different a priori roles. Kinds enter conceptual connections only insofar as they are presented under concepts. Hence, if we want to claim that type-identity theory is a metaphysical position (and this is a substantial point, not a verbal point about how we should use terms such as 'concept', 'kind', or 'type), then we should not talk as if the objects this theory is concerned with were the possible objects of procedures such as *conceptual analysis*. This is captured by the explication offered in Chap. 5: On this view, conceptual issues are raised on the level of representational devices rather than on the level of what is represented. In the framework defended here: Conceptual relations are mirrored by relations between property structures. Relations among the objects that are determined by property structures mirror the metaphysical connections relevant for identity-theories of the mind – if water reduces to H_2O, then the fact that water $= H_2O$ is, in a sense, a fact about what is represented by the conceptual contents of 'water' and 'H_2O', whereas conceptual differences between 'water' and 'H_2O' are captured by the corresponding property structures.

Now, type identity theories have been extensively criticized. Physicalists opposed to type-identity theory usually conceive of themselves as *non-reductive* physicalists. Interestingly, understanding what non-reductive physicalism consists in requires us to understand what this doctrine is opposed to – namely, reductionism about (mental) types. Moreover, non-reductive physicalists often describe themselves as *token-identity theorists*: they assume that mental tokens *reduce* to physical tokens, so that there is strong unity at the token-level. Again, a notion of identity-based reduction is required to fully understand this idea. The explication offered here cannot only be fruitfully applied in the context of type identity theory. The next section is dedicated to the connection between the explication of reduction and token identity theories.

6.4 Token Identity Theory

In a first approximation, the most prominent kind of argument[3] raised against type identity theory is this: Higher-level types can be realized in many different ways. That is: There are a number of physical or basic kinds the instances of which

object it represents. Talking about *concepts* in this fashion does not contribute anything substantial to the ontological talk, it merely enriches our vocabulary.

[3]There are many different versions of the argument we cannot go through here. The most relevant versions can apparently be found in (Putnam 1967; Fodor 1974, 1997). For detailed discussions, see (Lewis 1969; Endicott 1993; Kim 1989, 1992; Shapiro 2000).

play the role prescribed by higher-level types. On this basis, the idea of cross-level identification of kinds is dismissed. On this view a given kind, like pain, can occur in many different forms: Human pain may be vastly different from Octopus-pain, and even more radically different from pain of possible silicon based life-forms – not phenomenally, but physically or chemically (on the critical side: Putnam 1967; Fodor 1974; Block 1978; Endicott 1993; on the side of those who tried to accommodate multiple realizability with reductionism: Churchland 1986, ch.7; Lewis 1969, 1972; Kim 1989, 1992). Thus, pain is multiply realized, or so we are invited to assume. Intuitively: Pain is multiply realized in the sense that there are many different (possible or actual) ways of being a pain – in the non-phenomenal reading. Realization is spelled out in different ways. Intuitively, we can think of it in terms of the relation holding between *determinables* and *determinates* (Yablo 1992): A pair of realized and realizing property is similar to a pair of determinable and determinate property. Shoemaker (2001) argues that the realized property's causal powers form a subset of the causal powers of the realizing property. Note that both these criteria can only be met if for any instance of a realizing property, P, it is an instance of the properties P realizes. For what follows, I shall thus stick to the intuitive understanding of a necessary condition for *multiple realization at some world*: A kind K is multiply realized by the elements of a set $\{K^*\}$ of kinds in a world w only if in w, any instance of any element of $\{K^*\}$ instantiates K. We should add that $K \notin \{K^*\}$, and that $\{K^*\}$ has at least two elements.[4] This will suffice for our present concerns; a slightly more detailed characterization of the argument will be given in Sect. 6.5.

This book is not about reductionism and, accordingly, not about the question of whether or not arguments from multiple realizability affect the truth of type-identity theory. However, this book assumes that the concept of reduction is *relevant*. If the argument from multiple realizability goes through, and if token-identity theories are in some sense non-reductive, then one might wonder whether the concept of reduction is relevant. Let me thus show that token-identity theories employ a notion of reduction similar to the one employed by type-identity theories – a notion that can, again, be defined based on the explication offered above.

Intuitively, token-identity theory can be described in analogy to type-identity theory: For any token of any mental kind, it is identical to a physical token. Note firstly that this version of token-identity theory follows from type-identity theory

[4]We get a necessary condition for another predicate, namely, '_is multiply realized across worlds w and w*' by giving the set of realizing kinds a different interpretation: it now comprises kinds of w and w*, and the multiply realized kind is instantiated by instances of different elements of this set at w and w*. Multiple realizability is then such that there are at least two possible worlds in which the kind is multiply realized.

It is noteworthy that the arguments from multiple realizability did not take a purely epistemological form, namely, that the epistemic possibility of multiple realizability will make it impossible to uncover and describe the chemical, biological or physical kind we have to identify with the mental one, but rather came in a metaphysical shape: In virtue of being multiply realized, mental kinds do not belong to the set of lower-level kinds.

as formulated above. Thus, we should keep in mind that to produce an appropriate token-identity theory, we have to add the denial of type-identity theory. Moreover, token-identity theory then is a clear case of *property dualism* (at least). It commits its supporters to the claim that everything there is in space and time is physical, even though the physical instantiates mental properties. Thus, the question of "whether the mental and the physical can in some sense be identified" (Feigl 1967, 71) is, again, answered positively, though it is given a different interpretation. Here, it is claimed that for identification, token-identity is sufficient.[5]

So, how does token-identity theory connect to the notion of reduction? Even though it is not a very common term in the literature, we sometimes find variants of the expression 'token reduction' (Cartwright 1999, 32 ff., Hooker 1981, part III; Bickle 1998, 223 ff.). Consider the following claim to get an idea of how token-identity theory (as intuitively sketched) relates to our paradigm case of reduction:

(i) Water is H_2O.

In the spirit of the adherents of type-identity theory, we would read this statement as follows:

(ii) (The kind) water is identical to (the kind) H_2O

Token-identity theorists (concerning folk-chemistry) may deny this claim and suggest a different reading:

(iii) Whatever has the property of being water has the property of being H_2O, and *vice versa*. (Presumably, there will be some modal operator governing this sentence.)

(iii) supports (iv):

(iv) This particular amount of water is identical to this particular amount of H_2O-molecules.

Does an explication along these lines, plus a denial of type-identity theory, capture the intuition behind token-identity theory? Token identity theorists assume that H_2O is more fundamental than water. Even if, on their view, the kind water does not reduce to the kind H_2O, there is directionality involved: *being a water-token* is less fundamental than *being an H_2O-token*. Similarly, if substance dualism is false then thoughts *reduce* to brain-processes, what appears to you as the referent of 'I', when uttered by you, *reduces* to the spatio-temporal, physical object or chain of events that is you, and your particular mind *reduces* to the set of particular neural and, maybe, bodily events, just like every other mind *reduces* to the corresponding neural and bodily substance or chain of events. Thus construed,

[5] Just as an aside: Nominalists should be cautious when embracing token physicalism as opposed to type-identity theory. Under a nominalist conception of *abstracta*, it may, depending on how modal operators enter the formulation of token identity theory, *imply* type-identity theory for co-extensional properties.

token-identity-theory is connected to some sort of object monism and reductionism concerning the level of *tokens* of kinds. Again, token-identity theory is designed to *reconcile diversity and directionality with strong unity at the token-level.* The mere identification does not account for diversity and directionality. Token-identity theory is underdetermined in one important respect. Building on the explication proposed above, we get a straightforward characterization of *token*-reduction:

(Explication VIII – Token-Reduction) *x, when presented under PS1,* ***token-reduces*** *to x, when presented under PS2 iff*

(i) *x is a token &*
(ii) *PS2 is a constitutive structure of x with respect to PS1.*

We could introduce the additional condition that the property structures involved here should be property structures that mimic the structure of multiple realization, reflecting, for example, the intuition that every mental token (that is: one object presented as an instance of mental kind *M*) reduces to a physical token (that is: the same object presented as instantiating a physical kind, where this kind is one of the realizers of *M*). This nicely fits the idea that the *realization relation* is an explanatory relation; to give an example, on this view, a pain token is a pain token *in virtue of* being a token of C-fiber stimulation.

The notions of type- and token-reduction introduced here bear upon an appropriate understanding of anti-reductionist aspirations in the context of ontological non-reductive physicalism: If we assume that there are different, irreducible kinds of properties, but still assume that token- or substance monism is true, then we could use the notion of token-reduction and type-reduction to capture this idea. Non-reductive physicalism, in its ontological version, consists in the affirmation of *token-reduction* and the denial of *type-reduction* for the relevant class of tokens and types. The explication proposed above enables us to give an idea of what *reductive* type- and token-identity theories consist in, and, combing the two, it yields a characterization of non-reductive physicalism. The next sections discuss the application of the explication of reduction to more recent versions of type-identity theory.

6.5 Type-Identity Theory Strikes Back

Does multiple realizability really attack type-identity theory? It seems obvious that many innocent properties are multiply realizable. For example, a specific mean kinetic energy of an entity can be realized by infinitely many spatial states of the entity's constituents (Kistler 1999). This does not affect its being a basic kind. Thus, multiple realizability does not seem to threaten identity-theories *per se*. The trick is to introduce a distinction between *heterogeneous* multiple realization and *homogeneous* multiple realization. The former is bad for the identity-theorist, whereas the latter is good. Let us grant that in fact, every high-level kind is multiply realized by heterogeneous lower-level types, such that cross-level identification is blocked in virtue of the heterogeneity of the realization-base. Thus, we grant the truth of the following argument's first premise:

(i) Higher-level types are multiply realized by heterogeneous physical types.
(ii) If some type is possibly realized by the heterogeneous physical types, then there is no physical type to which it is identical.
(iii) Therefore higher-level types are not to be identified with physical types.[6]

As already mentioned, the argument comes in different forms – some tell the story in counterfactual terms (Lewis 1980), some claim that actually, mental kinds are multiply realized (Kim 1992; Fodor 1974; Block and Fodor 1972; Putnam 1967; Endicott 1993) – and it should be noted that type-identity theory is lost even if there were just *one* high-level type which is multiply realized, if the argument is sound.

In its classical form, the argument from multiple realizability remains silent about criteria for innocent and problematic versions of heterogeneity of the realizers, and about good candidates for kinds that are to be identified with higher-level types. Additional considerations have to be taken into account, for example considerations concerning conditions on when a homogeneous type is describable in disjunctive terms (Kim 1992). In recent years, reductionists have tried to argue that in some sense or another, the truth of the claim that high-level properties are multiply realized by lower level ones is compatible with reductionism which is based on cross-level identification of types. There are mainly four strategies used by reductionists to face the threat of multiple realizability. Some have tried to show that the kinds realizing high-level kinds are not as heterogeneous as the anti-reductionist wants us to believe, such that, for example, reductionism concerning the mental is not threatened by multiple realizability, just because the relevant high-level kinds are not multiply realized in any problematic sense (see, for two representative strategies: Shapiro (2000); Bechtel and Mundale (1999); The arguments take quite different forms: Bechtel and Mundale focus on actual examples, Shapiro raises more principled worries). In response, others have argued that mental kinds are not only multiply realized across actual and merely possible species (Kim 1992; Fodor 1974; Block and Fodor 1972; Lewis 1980; Putnam 1967), but also across individuals (Endicott 1993). The plasticity of the brain (Johnson et al. 1993), for example, gives reason to assume that quite different brain-areas can adapt to changes in other areas by taking over their function. So, given these problems, what are the prospects for property reduction? There remain three strategies to reconcile

[6]Note that this interpretation does not cover one specific version of token-identity theory. Some would argue that there just are no high-level types (or no mental types). One might want to argue that mental predicates just do *not* signify kinds (Kim 1998). This strong claim will be ignored for what follows. Thus, it is presupposed that mental predicates signify *kinds*. Early token-identity theorists (like Fodor), should be regarded as being happy with the assumption that some mental predicates signify mental kinds.

reduction with multiple realizability: by introducing *contextualized kinds* (6.5.1), by introducing *disjunctive kinds* (6.5.2) and by arguing- that *identity just is not required for reduction* 6.5.3).

6.5.1 Contextualized Kinds

One idea of how to account for reductions within a world in which high-level kinds are multiply realized is straightforward: Kinds are multiply realized in virtue of the fact that different physical kinds play their role on different occasions. Thus, if we individuate the high-level kinds more fine-grained, then we will be able to accommodate type-identity theory within a theory of multiple realizability: Even though kind k might be multiply realized by kinds k_1, k_2, ..., there is a context c_1, c_2 ... for any k_1, k_2, ... such that in c_1, k_1 plays the role of k, and in c_2, k_2 plays the role of k and so forth. The context could be species-specific, individual-specific or even specific with respect to different states within one individual. Thus, we could construe c-variants of k ('k in c_1' thus being an expression picking out a kind different from the kind picked out by 'k'), namely: k in c_1, k in c_2 and so forth. k in c_1 will thus be identical to k_1, k in c_2 will be identical to k_2 ... etc. To give an example: Octopus pain might differ from human pain, but this does not imply that human pain cannot be identified with a physical kind (namely, the kind which realizes pain in humans), and that octopus-pain cannot be identified with another physical kind (namely: the kind which realizes pain in octopuses). This idea goes back to David Lewis (1969, 1972), and it was further developed by Kim (dubbed: *local reductions*, Kim 1992), and Jackson et al. (1982). Esfeld and Sachse focus on reductions of sub-types of the high-level kind, so that contextualization occurs at the reduced level (Sachse 2007; Esfeld and Sachse 2007, 2011). Brian Loar (1981, ch. 4) suggests type-type identification for mental kinds relativized to times and individuals, where mental kinds are interpreted as first-order properties.[7] How does this strategy connect to the explication proposed above? If contextualization is supposed to lead to identification, then it is situated at the descriptive, or conceptual level. Classical type identity theory assumed that, say, pain = C-fiber stimulation. A contextualized version relativizes pain, say, to humans; on this view, pain in humans = C-fiber stimulation. Contextualization is supposed to ensure identities. But there is more than a verbal difference between 'pain in humans' and 'C-fiber stimulation'. Whereas the designatum of 'pain in humans' is supposed to be identical to the designatum of 'C-fiber stimulation', the

[7]Note that this poses another difficulty: Sometimes, it is argued that functional kinds (those that are multiply realized) are *second order*. For a criticism of this view, see my (van Riel 2012). In this paper, it is argued that there is no metaphysically relevant interpretation of being second order that captures the functionalist's intuitions about functional properties, and that there is in principle no reason to believe that functional properties are second order in virtue of being functional (in an interesting sense).

conceptual contents, and, correspondingly, the property structures they give access to are vastly different. Contextualization, at least in the form envisaged by those who defend contextualization as a move to save type-identity theory from the threat of multiple realizability, clearly does not give us a better insight into the object's nature, or access to a constitutive property structure. Rather, these contextualizations with respect to species, individuals or even times build on, in this respect, more or less arbitrary relational properties in order to specify the mental kind that is at issue, the kind that is to be identified with a physiological kind. Contextualized descriptions give access to property structures that do the job of *individuating* the right kind that can be identified with a neural or physiological or physical kind, but they do so in a way that is intransparent with respect to this kind's nature. Given the explication of reduction proposed above, contextualization and identification in this way *should* lead to reduction, because the neural or physiological descriptions are obviously more transparent to a physiological or neural kind's nature than their counterparts that involve contextual features (granted that naturalism is true). Appropriate contextualization leads not only to identification, but also to reduction. Human pain is not only identical to C-fiber stimulation; it also reduces to C-fiber stimulation, and not *vice versa*.

If identification of contextualized kinds is all that we possibly get, we have to deny property monism (non-contextualized high-level kinds are *not* physical) or, as is sometimes argued, we have to accept that predicates and terms allegedly picking out high-level kinds do not pick out kinds at all (this is Kim's strategy (e.g. Kim 2002)) – again an idea that can be nicely modeled within the present framework. On this latter view, we do not get identification because the property structures do not determine any entity at all. No property corresponds to the property structure – hence, we do not get identification. All we get is the property structure itself. Not so for the contextualized descriptions. When presented under a contextualized description, a kind x reduced to x when presented under a description that is more transparent to the properties in virtue of which it behaves the way it does, or is what it is. Directionality and diversity, as cashed out above, are, thus, constitutive for the contextualized kind approach.

6.5.2 Disjunctive Properties

Let us turn to the second option to save type-identity theory from the threat of multiple realizability. Jaegwon Kim famously argued in favor of local reductions (Kim 1992). One branch of his argument is designed to attack anti-reductionists like Fodor who hold that high-level kinds play a relevant role in natural laws. The strategy Kim employs is this: He argues that multiple realizability depends on there being different realizing properties. If there are different realizing properties, then any kind that is multiply realized is disjunctive in nature. A disjunctive kind is (intuitively speaking) referred to by disjunctive expressions the disjuncts of which are terms referring to basic kinds. Thus, disjunctive kinds are candidates for cross-level identification. However, Kim argues that often, the high-level kinds' tokens do not share a set of (scientifically) relevant features. The example he uses to illustrate this point is Jade. Jade is a disjunctive kind in the sense that x is jade iff x is either jadeite or nephrite. This, or so Kim argues, holds for any multiply realizable kind. What is this supposed to show? It is supposed to show that disjunctive kinds do not and should not figure in natural laws (not even in special science laws).[8]

Others have reacted in a different way. Clapp (2001) and Walter (2006) argued that the disjunctive move is legitimate: The high-level type is identical to some kind we can refer to by a term containing a construction of a number of disjuncts that designate low-level kinds (we can refer at least in principle, because the disjunction might be infinitely long). There are two questions to be distinguished here: Whether or not multiply realized kinds possibly figure as autonomous in high-level sciences (this is Fodor's target (Fodor 1997)), and whether or not disjunctive kinds can be identified with high-level kinds and, thus, whether or not type-identity theory is vindicated by an appropriate disjunctive move (Clapp 2001; Walter 2006). The *relata* in such cross-level identifications, which are based on disjunctive descriptions of the higher-level kind, are, clearly, types (on the assumption that the disjunctive strategy succeeds). Identification with disjunctive types corresponds to a reductive strategy. Directionality and diversity, as cashed out above, are, again constitutive for this approach: We get diversity in conceptual representation (disjunctive and non-disjunctive), and directionality that is intuitively tied to the reduced entity's nature: If the reductionist is right, the disjuncts of the disjunctive description exhaust the nature of the high-level type. Again, reduction reconciles diversity and directionality with strong unity. If we phrase this account in terms of identity alone, it turns out to be underdetermined with respect to diversity and directionality. Within the framework outlined in Chap. 5, the disjunctive move can be reconstructed as follows (given that it actually yields the required cross-level identifications): We could, at least in principle, give a disjunctive description of a mental property or kind, thereby presenting it under a disjunctive property structure. On this view, we get reduction only if the nature of the high-level kind is exhausted by the disjunction, that is, only if there is no explanation of why a mental phenomenon of kind k occurs that cannot be accounted for in terms of one of the disjuncts of the disjunction that captures the neural kind k is identical to. This seems to be the desired result: Mere co-intensionality is not enough to show that the physiological or neural level is

[8]One branch of the argument is based on the assumption that high-level kinds are not projectible. Let me just point to some fundamental confusion this debate seems to be based on: According to the classical interpretation, projectibility is a property of *predicates* (Goodman 1979; Quine 1969). As Antony (2008) conclusively argued: even if *predicates* (or, as we should add, maybe terms which designate a property) are not projectible, *properties* so designated might nevertheless be nomic properties. This observation perfectly matches the intuition that *being disjunctive* is, in a primary sense, a property a property has under some way of linguistic representation, namely, a disjunctive form of representation. Thus, it is hard to see how properties should be disjunctive *simpliciter*, and, thus, whether or not being disjunctive is relevant for the arguments against type-identification based on disjunctive kinds.

more fundamental than the mental level. Hence, the disjunctive move amounts to reduction only if the disjunctive description gives access to a property structure that is more transparent with respect to the kinds nature than the property structure the mental description gives access to. The feature of being disjunctive is reflected on the linguistic and the conceptual level; it is not reflected on the level of designation. For multiply realized kinds, the disjunctive characterization is obviously more transparent with respect to the kind's nature – something is jade, if it is, in virtue of being jadeite, or in virtue of being nephrite, and not *vice versa*. Again, this nicely fits the idea that diversity is conceptual in nature, whereas unity requires identity. Thus, the disjunctive move can be reconstructed in terms of the explication of reduction proposed here. If so, we come to see why it is inherently reductive, and not merely a strategy to obtain identifications, which remains neutral as to directionality and diversity. Directionality and diversity, as described above, are, thus, constitutive for the disjunctive move.

6.5.3 Alternatives to Identity

As already pointed out in the introduction, unity may come in degrees. There are weaker forms of unity that build upon notions weaker than that of identity. These have played a prominent role in the reduction debate in the philosophy of mind and should, thus, be mentioned briefly. Supervenience (Kim 1982, 1984, 1985, 1987) could be a serious candidate. Any contextualized-kind approach to reductionism could be extended to a general reductionist claim that connects some realization base with the relevant supervenient high-level kinds. On some occasions, Kim discusses reduction via supervenience (1984). It should be noted, however, that according to any notion of supervenience, in a straightforward sense, what supervenes or is realized is not (necessarily) identical to its supervenience-, or realization-base.

Choosing this strategy is interesting when we are concerned with issues regarding the appropriateness of high-level sciences within some weaker form of physicalism than strict monism. It should be noted that once we take this path, we give up on trying to reconcile diversity and directionality with strong unity in the sense specified above. To deny that we live in a strictly monistic world does not give an answer to the question of how, in a strictly monistic world, there can be a diversity of theories and, intuitively, levels we describe in different sciences and in every-day life – it is to deny that this question can be answered in an interesting way for the actual world. If identity is replaced by a weaker relation, we weaken the monist's claim: According to this view, there are some kinds which are not identical to fundamental kinds. Thus, if we were to apply these strategies to our explication of the notion of reduction, we would change the subject.

6.6 An Intermediate Conclusion: *Identity*-Theories

The assumption underlying a large part of the reduction debate in the philosophy of mind is that identification of types is a prerequisite for successful reduction. This implicit assumption underlies the idea that, even if token-identity theory is true, identification of contextualized-types is still possible. Alternatively, it might be the case that type-identity theory is rescued by identification of mental types with disjunctive types. Finally, it might turn out that we should give up type-identity theory altogether, namely, when we come to see that it is not available due to the multiple realizability or similarly problematic features of high level kinds (this is the strategy of non-reductive physicalism, which comes in a great variety: *cf.* van Gulick 1992; Cartwright 1999; Davidson 1970, 1993; Fodor 1974, 1997; Putnam 1967). Even if type-identity theory is given up, we still have token-identity theory, sometimes conceived of as a new version of substance monism and as a prominent form of non-reductive physicalism. On both assumptions, type- and token-identity theory, we might be interested in how a monistic world (on the type- as well as on the token-level) can be accounted for using different every-day descriptions and different sciences and theories. Thus, on both interpretations, it is worth answering the question of reduction: How can we reconcile diversity and directionality with strong unity? It has been shown that the explication proposed here can be used to reconstruct various forms of type-identity theory, token-identity theory and non-reductive physicalism. When type identity theory is based on contextualization or the disjunctive move, the notion of reduction explicated above is built into the theory right from the start.

Up to now, we have reflected upon a family of partly competing and partly complementary theses about the metaphysics of the mind, and on how these theses can be illustrated by reference to the explication of reduction. These theses remain silent about the epistemic or procedural character of reductions (except for the highly implausible idea that the relevant identity-statements express a priori truths) and about explanatory aspects of reduction. In the philosophy of mind, the dominant view about the link between reduction and explanation is that *functional* reduction somehow goes together with functional or mechanistic explanation.[9] Functionalism also yields a theory of the metaphysics of mental properties and, contrary to ordinary type-identity theories, it explicitly employs the term ‘reduction’.

[9]These notions are left intuitive (see Block (forthcoming) for an exception). Sometimes, reference to the Nagelian idea of reduction as explanation can be found in the philosophy of mind debate (cf. Kim 1993; Fodor 1975).

6.7 Functional Reduction

Among philosophers of mind, it is common to assume that at least some mental properties are functional in nature (Putnam 1975; Jacob 1997; Kim 1992, 1998; van Gulick 2001; Block 1997, 2007, forthcoming). Basically, a functional property is a property that is to be characterized, or that is characterizable in functional terms. The idea that functional properties are functional in *nature*, or that they are *role-properties* raises subtle issues about the nature of dispositional properties, their connection to multiple realizability and to functionalism. Suffice it to say that on all these views, a functional characterization is, to some extent, *transparent* with respect to the so described property's nature (for a discussion, see van Riel 2012). Here are two ways to conceive of characterizations in functional terms. The first is obtained from Lewis (1972), the second can be found in Block (2007), building on Putnam (1975). To give an idea of how to conceive of functional properties, it will prove useful to talk freely about *theoretical* terms. Theoretical terms are conceived of as terms that can be defined by the Ramsey-sentence of the theory in which they occur. A functional characterization of a property is, then, a characterization of a property in terms of the relevant Ramsey-sentence, obtained from a psychological theory.

The first idea underlies Lewis' proposal to describe functional reduction. On Lewis' interpretation (Lewis 1972), from the statement of a psychological (folk-) theory, schematically: T[t], with theoretical (in our case: mental) terms t, we can obtain a modified Ramsey-sentence by replacing the theoretical terms by variables within the scope of existential (uniquely quantifying) quantifiers, schematically: $\exists_1 x$ T[x] (with $\exists_1 x$ F$x \leftrightarrow \exists x$, $\forall y$, (Fx & F$y \rightarrow y = x$)). From this, we can construct a characterization of the mental terms occurring in this theory. To give an example, which is in the spirit of Lewis, we can characterize a singular mental term 'a' as follows, ignoring unique quantification ('i_1', 'i_2' ... stand for input-state terms and 'o_1', 'o_2' ... for output-state terms):

> $a =$ the x which is such that $\exists y$, z, ... (under i_1, x causes y and o_1, under i_2 and given z, x causes o_2 ...).

Here, we have characterized the referent and, maybe, the meaning of a referential term. The referent is the property described by the modified Ramsey-sentence on the right-hand side of the identity sign. Ned Block relies on an alternative interpretation, focusing on definitions of predicates or open sentences rather than characterizations of terms. Block writes:

> To get the Ramsey sentence of T, replace the mental-state terms [...] by variables, and prefix an existential quantifier for each variable: $\exists F_1 \ldots \exists F_n$ T($F_1 \ldots F_n$, $I_1 \ldots I_k$, $O_1 \ldots O_m$). If F_{17} is the variable that replaced 'pain' when the Ramsey sentence was formed, we can define pain as follows in terms of the Ramsey sentence: x is in pain $\leftrightarrow \exists$F1 ... $\exists$Fn T[($F_1 \ldots F_n$, $I_1 \ldots I_k$, $O_1 \ldots O_m$) & x has F_{17}] (Block 2007, 68)

Here, a *predicate* is defined (and the Ramsey-sentence schema is slightly changed by adding the schematic expression '& x has F_{17}'). In contrast, we gave a characterization of the referent of a *name* above. Now, both these sorts of characterizations can be conceived of as functional characterizations of a property. Assuming that mental or other properties are functional, it will be assumed, in this section, that mental or other properties can (or have to) be characterized in one of these ways.

Building on this notion of functionalism, Lewis (1970, 1972), Armstrong (1968) Chalmers (1996) and Kim (1992, 2006) developed a model of reduction, which is largely independent of a decision on the issue of *multiple* realizability, but has come to be the dominant view on reduction in the philosophy of mind and in many areas in the philosophy of science. According to functionalism, a theory contains theoretical terms. Deleting these terms, we get an open sentence that can, with slight syntactic modifications, be used to define the property the deleted theoretical terms allegedly refer to. This predicate is functional in the following sense: it makes the *functional role* of that property transparent. It thereby gives us the relevant information to enable us to find a property that satisfies this description. Assume that pain is nothing but C-fiber firing. If pain is construed functionally, by generating a predicate from folk-psychology by deleting 'pain', we get a description that is satisfied by C-fiber firing. C-fiber firing plays the role specified by the open sentence. Now, the idea is that therefore, and in this very sense, pain reduces to C-fiber firing.

Models of functional reduction come in two epistemological forms: Some functionalists hold that arriving at a functional definition is by and large an a priori task. Others assume that it is not; it is an empirical matter *which* functional role the alleged referents of our theoretical terms play. Intuitively, functional reductionism is the doctrine that there is a procedure (a priori or empirical) to find out about cross-level identities, which is justified by the functional nature of the item to be reduced. On this conception, functional reduction remains silent about what grounds the directionality, and it merely implicitly suggests what the diversity consists in (as we will see below, it is, again, conceptual in nature). Definitions of functional reduction sit, so to speak, on top of more robust definitions of reduction, such as Nagelian reduction, according to which reduction is derivation of one theory from another with the help of bridge-laws. Functionalism in itself has got nothing to say about reduction in this sense. This can easily be seen: Functional reduction is sometimes said to be (and, *prima facie* is) compatible with (at least) *supervenience* of functional properties on lower-level properties as well as with *identity* (Lewis 1972). Non-reductive physicalists as well as identity-theorists can be functionalists. They do not disagree about the fact that some properties are functional in nature. They disagree about how this connects to reduction. A functionalist version of Nagelian reduction has been defended by Esfeld and Sachse, as noted above: According to Esfeld and Sachse, a theory may reduce to another theory by identification of functional sub-types of the reduced theory with types of the reducing theory, thereby getting the required bridge-laws (Esfeld and Sachse 2007). Thus, functionalism is open to further interpretations of the ontological link. Functional *reductionism*, in addition, assumes that functional analysis tracks identities.

So, roughly, functional reduction is a procedure that is based on (i) investigation of *semantic* (in it's a priori form) or *empirical* (in it's a posteriori form) links

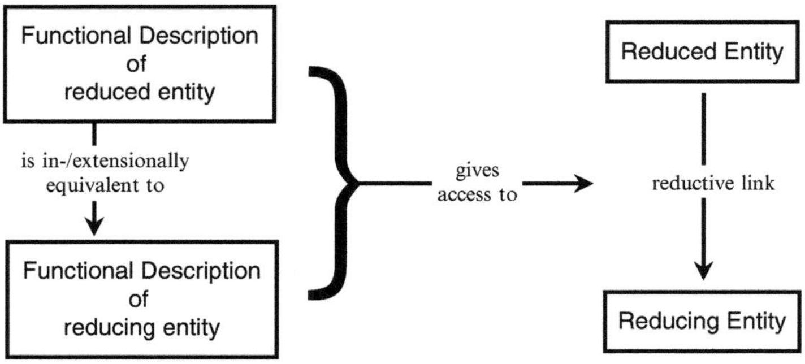

Fig. 6.1 Functional reduction

that give the resources to (ii) establish empirical links, like finding the relevant realizers, which in turn, according to some functionalists (iii) justify a reductionist claim. The epistemic insight to be gained from this procedure is this: We find the physical property which plays the functional role described by the theory by functionally characterizing the relevant property (Kim 1992, 2006; Lewis 1972; Levine 1993, 132). The first step (which makes it *functional*) consists of conceptual analysis or empirically inspired reconstruction in functional terms. The functional characterization makes the causal role of the property transparent. Thereby, a functional "analysis" enables us to look for another description that describes the actual occupiers (the elements of the concept's extension in the actual world) of the relevant functional role in different terms. We are then in a position to justify our claims concerning the relevant links between the *relata* of the reduction relation. Figure 6.1 illustrates this point (I choose a simple version just mentioning straightforward reductions and ignoring mere replacements as well as reduction-relations based on supervenience of the reduced and the reducing entity):

This sketch captures the idea that functionalism concern the *procedure* of reduction, not the reductive link itself. We use a functional description of the reduced item and a functional description of the reducing item to empirically investigate their interrelations. When we then discover a reductive link (in the case relevant for us: identity) we can move to the claim that in fact, the candidate item actually reduces to the reducing item.

Let us assume that the model of functional reduction is not a purely empirical model that is supposed to capture how we actually carry out reductions. Nor is it a purely normative model that is concerned with how we should carry out reductions. Rather, it should be interpreted as a partial definition. If so, this approach is committed to the claim that at least some of the *relata* of the reduction relation are *concepts* (or modes of presentations, or senses of the expressions which refer to the items which are said to reduce), because it is the *descriptive content* of the terms that makes for diversity. The relevant *concepts* present us with an item, a *functional property*, which is then said to be identical with a type at another level. It is easy to see how this relates to reduction as conceived of here:

Functional properties are described in a canonical form, namely, in terms of how they relate to input and output states. This nicely fits the closely related assumption that functional reduction is somehow tied to "functional" explanation. Let us, again, stipulate that a person's being in pain is its being the subject to specific C-fiber stimulation. Intuitively, asking *how pains occur in humans*, or *what it is for a human to be in pain*, or even *why a human feels, or is in pain* we can respond that pains occur by C-fiber stimulation, or that for a human to be in pain is for her to be the subject of specific C-fiber stimulation, or that she is in pain *because* she is the subject of C-fiber stimulation. To make this idea more precise, consider how a re-description of a functional mental property in physiological terms gives rise to the explanatory link. Functional characterizations characterize *mental* properties by relating them to behavioral and other internal states. Physiological characterizations of the very same properties yield the resources to *explain why the mental property occupies the causal role that is explicit in the functional characterization we started with.* Assume again that being in pain is identical to being caused by tissue damage and causing yelling. Assume that being in pain is identical to being a C-fiber firing. Therefore, being a C-fiber firing is identical to the complex property (or the property designated by the complex description) of being caused by tissue damage and causing yelling. If the reductionist is correct, something causes a yelling when caused by tissue damage *because* it is a C-fiber firing. This is the idea of functional reduction. By finding the entities that play the relevant functional role, and by describing them at a lower level of description, we come to see why the causal connections transparent in the functional description we started with obtain. A mental object satisfies the property structure of a functional description that relates the object to inputs and bodily outputs, if it does, *because* it satisfies the property structure made available at the physiological level of description. Given the model of reduction explicated above, we come to see how functional explanation and functional characterizations of a property interconnect. This is captured by the reconstruction of functional reduction suggested above.

An individual functionalist reduction statement about types can be captured as follows:

(Functional Type-Reduction) *x functionally reduces to y iff*

(i) *x, when presented under PS1,* ***type-reduces*** *to x, when presented under PS2, &*
(ii) *PS1 and PS2 describe x in functional terms.*

A similar notion of token reduction and, thus, of non-reductive versions of functionalism can easily be defined: For token-reduction we just replace 'type' by 'token' in condition (i), and for non-reductive versions, we combine the denial of *Functional Token-Reduction* with *Functional Type-Reduction.* These characterizations also illustrate in which sense models of functional reduction are parasitic on other notions of reduction; whether or not a model of reduction is functionalist depends on additional features of the model.

6.8 Conclusion

The explication proposed in the first part of this book sheds light on conceptions in the philosophy of mind, such as various conceptions of type identity theory, token identity theory and non-reductive physicalism. We thereby motivated the explication – it is not only adequate, but also fruitful.

Q2: How can this explication be further motivated?
Th. 5: It sheds light on the reduction debate in the philosophy of mind.
...

We have seen that most of the strategies pursued in attacks against anti-reductionism share one common feature: They try to show that cross-level identities are available. However, they differ in degree of strength: Some argue in favor of a weaker form of compatibility of multiple realizability and reductionism, others argue in favor of a stronger form of compatibility. Token as well as various forms of type-reductionism rely on a notion of reduction as based on cross-level identities. One alleged epistemic virtue of such identifications has not been commented on yet. Block and Stalnaker consider the epistemic status of a posteriori identity statements (Block and Stalnaker 1999, 23n). Basically, they argue that we come to believe that kind k_1 and kind k_2 are identical rather than correlated by inference to the best explanation. During their discussion they mention an interesting epistemological distinction between identity statements and statements of a mere correlation, which gives reason to assume that reductionist aspirations are based on seemingly fundamental and fairly general epistemic principles:

> If we believe that heat is correlated with but not identical to molecular kinetic energy, we should regard as legitimate the question of why the correlation exists and what its mechanism is. But once we realize that heat is mean kinetic energy, questions like this will be seen wrongheaded. (Block and Stalnaker 1999, 24)

If we want to follow Block and Stalnaker, we have an additional reason to regard a notion of reduction as being based on identity to be extremely important: Only if we regard bridge-principles (or whatever does the linking-job) to state identities the reduction has succeeded, because otherwise, we would still be in a position to ask *what the underlying mechanism is* which is responsible for the correlation.[10]

The two most prominent ontological positions in the philosophy of mind, type-identity theory and token-identity theory, build upon an understanding of reduction that is captured by the job-description that reduction is supposed to reconcile diversity and directionality with strong unity. These theories differ in their interpretation of where to look for unity – on the level of types or on the level of tokens. As such, they do not differ in other respects. Both sorts of theories are underdetermined with respect to directionality and diversity. The explication proposed above fixes these deficiencies. Moreover, it neatly matches the commitments concerning the

[10]Note that this idea has been questioned (cf. Kim 2008); for a defense, see (van Riel 2010).

relevance of conceptual aspects in type-identity theories in the writings of Smart, in the context of the disjunctivist's as well as the contextualist's move and in the context of functionalist models of reduction. It can be used to define ontological non-reductive physicalism, type and token reduction as well as functional reduction, a theory that is explicitly committed to the idea that reduction relations are sensitive to descriptions. A plausible assumption about the connection between functional reduction and functional explanation is vindicated and explained.

Bibliography

Antony, Louise M. 2008. Multiple realization: Keeping it real. In *Being reduced. New essays on reduction, explanation and causation*, ed. Jesper Kallestrup and Jakob Hohwy, 164–175. Oxford: Oxford University Press.

Armstrong, David. 1968. *A materialist theory of the mind*. London: Routledge & K. Paul.

Bechtel, William, and Jennifer Mundale. 1999. Multiple realizability revisited: Linking cognitive and neural states. *Philosophy of Science* 66: 175–207.

Bickle, John. 1998. *Psychoneural reduction: The new wave*. Cambridge, MA: MIT Press.

Block, Ned. 1978. Troubles with functionalism. *Minnesota Studies in the Philosophy of Science* 9: 261–325.

Block, Ned. 1995. The Mind as the software of the brain. In *An invitation to cognitive science 3: Thinking*, ed. Daniel N. Osherson and Edward E. Smith, 377–426. Cambridge, MA: MIT Press.

Block, Ned. 1997. Anti-reductionism slaps back. *Philosophical Perspectives* 11: 107–132.

Block, Ned. 2007. *Consciousness, function, and representation*. Cambridge, MA: MIT Press.

Block, Ned. forthcoming. Functional reduction. In *Supervenience in mind: A Festschrift for Jaegwon Kim*, ed. Terry Horgan, David Sosa and Marcelo Sabates. Cambridge, MA: MIT Press.

Block, Ned, and Jerry A. Fodor. 1972. What psychological states are not. *Philosophical Review* 81: 159–181.

Block, Ned, and Robert Stalnaker. 1999. Conceptual analysis, dualism, and the explanatory gap. *Philosophical Review* 108: 1–46.

Cartwright, Nancy. 1999. *The dappled world: A study of the boundaries of science*. Cambridge: Cambridge University Press.

Chalmers, David. 1996. *The conscious mind*. Oxford: Oxford University Press.

Churchland, Patricia. 1986. *Neurophilosophy*. Cambridge, MA: MIT Press.

Clapp, Leonard. 2001. Disjunctive properties and multiple realizations. *The Journal of Philosophy* 98: 111–136.

Davidson, Donald. 1970. Events as particulars. *Noûs* 4: 25–32.

Davidson, Donald. 1993. Thinking causes. In *Mental causation*, ed. Alfred Mele and John Heil, 3–18. Oxford: Clarendon.

Endicott, Robert. 1993. Species-specific properties and more narrow reductive strategies. *Erkenntnis* 38: 303–321.

Esfeld, Michael, and Christian Sachse. 2007. Theory reduction by means of functional sub-types. *International Studies in the Philosophy of Sciences* 21: 1–17.

Esfeld, Michael, and Christian Sachse. 2011. *Conservative reductionism*. London: Routledge.

Feigl, Herbert. 1967. *The "mental" and the "physical". The essay and a postscript*. Minneapolis: University of Minnesota Press.

Fodor, Jerry A. 1974. Special sciences: Or the disunity of science as a working hypothesis. *Synthese* 28: 97–115.

Fodor, Jerry A. 1975. *The language of thought*. New York: Thomas Crowell.

Fodor, Jerry A. 1997. Special sciences: Still autonomous after all these years. *Noûs Supplement: Philosophical Perspectives 11, Mind, Causation, and World* 31: 149–163.
Frege, Gottlob. 1892. Über Sinn und Bedeutung. *Zeitschrift für Philosophie und philosophische Kritik* 100: 25–50.
Goodman, Nelson. 1979. *Fact fiction and forecast*. Cambridge, MA: Harvard University Press.
Hooker, Clifford. 1981. Towards a general theory of reduction. Part I: Historical and scientific setting. Part II: Identity in reduction. Part III: Cross-categorial reduction. *Dialogue* 20: 38–59, 201–236, 496–529.
Jackson, Frank. 2005. The case for a priori physicalism. In *Philosophy-science -scientific philosophy, main lectures and colloquia of GAP 5, fifth international Congress of the Society for Analytical Philosophy*, ed. Ansgar Beckermann and Christian Nimtz, 251–265. Frankfurt/M: Mentis.
Jackson, Frank, Robert Pargetter, and Elizabeth Prior. 1982. Functionalism and type-type identity theories. *Philosophical Studies* 42: 209–225.
Jacob, Pierre. 1997. *What minds can do. Intentionality in a non-intentional world*. Cambridge: Cambridge University Press.
Johnson, Mark H., Yuko Munakata, and Rick Gilmore. 1993. *Brain development and cognition*. Oxford: Blackwell Publishing.
Kim, Jaegwon. 1982. Psychophysical supervenience. *Philosophical Studies* 41: 51–70.
Kim, Jaegwon. 1984. Concepts of supervenience. *Philosophy and Phenomenological Research* 45: 153–176.
Kim, Jaegwon. 1985. Supervenience, determination, and reduction. *The Journal of Philosophy* 82: 616–618.
Kim, Jaegwon. 1987. 'Strong' and 'global' supervenience revisited. *Philosophy and Phenomenological Research* 48: 315–326.
Kim, Jaegwon. 1989. The myth of nonreductive materialism. *Proceedings and Addresses of the American Philosophical Association* 63: 31–47.
Kim, Jaegwon. 1992. Multiple realization and the metaphysics of reduction. *Philosophy and Phenomenological Research* 52: 1–26.
Kim, Jaegwon. 1993. *Supervenience and mind. Selected philosophical essays*. Cambridge: Cambridge University Press.
Kim, Jaegwon. 1998. *Mind in a physical world. An essay on the mind-body problem and mental causation*. Cambridge, MA: Harvard University Press.
Kim, Jaegwon. 2002. The layered model: Metaphysical considerations. *Philosophical Explorations* 5: 2–20.
Kim, Jaegwon. 2005. *Physicalism, or something near enough*. Princeton: Princeton University Press.
Kim, Jaegwon. 2006. *Philosophy of mind*. Boulder: Westview Press.
Kim, Jaegwon. 2008. Reduction and reductive explanation. Is one possible without the other? In *Being reduced. New essays on reduction, explanation and causation*, ed. Jesper Kallestrup and Jakob Hohwy, 93–114. Oxford: Oxford University Press.
Kistler, Max. 1999. Multiple realization, reduction and mental properties. *International Studies in the Philosophy of Science* 13: 135–149.
Levine, Joseph. 1993. On leaving out what it's like. In *Consciousness: Psychological and philosophical essays*, ed. G. Humphreys and M. Davies, 121–136. Oxford: Blackwell.
Lewis, David. 1969. Review of art, mind, and religion. *The Journal of Philosophy* 66: 23–35.
Lewis, David. 1970. How to define theoretical terms. *Journal of Philosophy* 67: 427–446.
Lewis, David. 1972. Psychophysical and theoretical identifications. *Australasian Journal of Philosophy* 3: 249–258.
Lewis, David. 1980. Mad pain and Martian pain. In *Readings in the philosophy of psychology*, ed. Ned Block, 216–222. Cambridge, MA: Harvard University Press.
Loar, Brian. 1981. *Mind and meaning*. Cambridge: Cambridge University Press.
Place, Ullin. 1956. Is consciousness a brain process. *British Journal of Psychology* 47: 44–50.
Place, Ullin. 1960. Materialism as a scientific hypothesis. *Philosophical Review* 69: 101–104.

Putnam, Hilary. 1967. Psychological predicates. In *Art, mind, and religion*, ed. D.D. Merrill and W.H. Capitan, 37–48. Pittsburgh: University of Pittsburgh Press.
Putnam, Hilary. 1975. On properties. In *Mathematics, matter and method. Philosophical papers*, ed. Hilary Putnam, 305–322. Cambridge: Cambridge University Press.
Quine, Willard V.O. 1969. *Ontological relativity and other essays*. New York: Columbia University Press.
Sachse, Christian. 2007. *Reductionism in the philosophy of science*. Frankfurt/M: Ontos-Verlag.
Shapiro, Lawrence A. 2000. Multiple realizations. *The Journal of Philosophy* 97: 635–654.
Shoemaker, Sidney. 2001. Introspection and phenomenal character. *Philosophical Topics* 28: 247–273.
Smart, John J.C. 1959. Sensations and brain processes. *Philosophical Review* 68: 141–156.
Smart, John J.C. 1963. Materialism. *Journal of Philosophy* 60: 651–662.
Thagard, Paul. 1990. The conceptual structure of the chemical revolution. *Philosophy of Science* 57: 183–209.
Van Gulick, Robert N. 1992. Nonreductive materialism and the nature of intertheoretical constraint. In *Emergence or reduction? Essays on the prospects of non-reductive physicalism*, ed. Ansgar Beckermann, Hans Flohr, and Jaegwon Kim, 157–179. Berlin: de Gruyter.
Van Gulick, Robert N. 2001. Reduction, emergence and other recent options on the mind/body problem: A philosophic overview. *Journal of Consciousness Studies* 8: 1–34.
van Riel, Raphael. 2010. Identity-based reduction and reductive explanation. *Philosophia Naturalis* 47–48: 183–219.
van Riel, Raphael. 2012. Pains, pills, and properties – Functionalism and the first-order/second-order distinction. *Dialectica* 66: 543–563.
Walter, Sven. 2006. Multiple realizability and reduction: A defense of the disjunctive move. *Metaphysica* 9: 43–65.
Yablo, Stephen. 1992. Mental causation. *Philosophical Review* 101: 245–280.

Chapter 7
Conceptions of Reduction in the Philosophy of Science

This chapter is about the notion of reduction in the philosophy of science. Going through the most influential (families of) characterizations of reduction, it is argued that a particular version of identity-based reduction plays a crucial role in the reduction debate in the philosophy of science. Again, it is a notion that reconciles diversity and directionality with strong unity, without relying on elimination. However, it is a notion the primary relata of which are *theories*. It is argued that models of reduction, in a straightforward sense of 'reduction', as well as models of mere replacement heavily build upon this notion. At the same time, the present section paves the way for an evaluation of the relation between versions of theory-reduction, or what will be called 'holistic' approaches to reduction, and the explication proposed in Chap. 5. In passing, it will be shown that models of reduction in the philosophy of science do not draw a sharp distinction between issues regarding replacement and issues regarding genuine reduction, and that *extension first criteria* (see Sects. 2.2 and 2.3) heavily influenced the discussion.

7.1 Identity Based Theory Reduction

Whereas in the philosophy of mind the notion of reduction is frequently treated as the basis for *ontological reductionism*, this is a thesis that is not of primary interest for most parts of current philosophy of science. The distinction between reduction debates in the philosophy of science and in the philosophy of mind rests on an idealization, but it nicely illustrates two different tendencies in the two fields. One of the main differences between the philosophy of science debate and the debate in the philosophy of mind becomes apparent when we read the very first sentence of Jerry Fodor's (1974) seminal paper:

> A typical thesis of positivistic Philosophy of Science is that all true theories in the special sciences should reduce to physical theories in the long run. (Fodor 1974, 97)

R. van Riel, *The Concept of Reduction*, Philosophical Studies Series 121,
DOI 10.1007/978-3-319-04162-9_7, © Springer International Publishing Switzerland 2014

For Fodor, 'reduction' is an expression that can be used to describe a certain metaphysical picture of the organization of sciences, just like it was for the early positivists. It is this picture which survived in the philosophy of mind but which was seemingly abandoned in the philosophy of science, although, as will be pointed out below, it still lurks within every single model of reduction on the market. This difference is reflected by the fact that in the philosophy of science, reduction is usually explicated in terms of a relation that primarily holds between *theories*, or representational devices, rather than non-representational worldly objects. Attempts to characterize the reduction-relation go back to the early years of logical positivism. Carnap's reductionism states that:

> ... science is a unity, [such] that all empirical statements can be expressed in a single language, all states of affairs are of one kind and are known by the same method. (Carnap 1934, 32)

This passage covers three dimensions of reductionism: A linguistic/semantic aspect (that all empirical statements can be expressed in a single language), an ontological aspect (that all states of affairs are of one kind) and a methodological aspect (that these states of affairs can be known by the same method). Translation is based on or should at least accord with a priori or analytic connections between different conceptual realms (what is referred to as "epistemological" reduction in Oppenheim and Putnam (1958, 3)). There must, then, be an analytic connection between the realm of the social sciences and that of physics, if there is at least one truth expressible in the language of the social sciences and if physicalism is correct. This optimistic idea was dismissed when it turned out that a priori reduction was beyond what we could actually hope for. As Hempel put it: "[T]he definitions in question could hardly be expected to be analytic" (1966, 103). Hempel suggested that, instead of looking for meaning-equivalence of statements in different sciences, we ought to conceive of the equivalence in terms of extensional equivalence (Hempel 1966, 104). When the Carnap-model was abandoned, two of the three aspects of reductionism survived in the context of reduction, namely, a more liberal variant of the linguistic aspect and the ontological aspect. The methodological aspect was dismissed just like the a priori aspect, and rightly so: It just is not the case that we know all truths by one method, and, moreover, it is hardly conceivable that there will be a time when we will be able to do so, even if we live in a monistic world. How to account for tax-raise with the methods of physics, even if tax raise just is a physical event – an admittedly extremely complex one, but still physical in nature? It seems impossible. But this is not to say that there are no physical events identical to tax-raise events. So, the ontological aspect and the methodological aspect *should* be kept separate.

Thus, from the beginning, the discussion of issues of reduction was tied to scientific unification, to the epistemology of theory relations, and to scientific practice. These issues became even more pressing when the orthodox view came under attack due to criticisms of Kuhn (1962) and Feyerabend (1966, 1981); attention shifted from seemingly straightforward reductions to questions regarding theory-change, replacement, and procedures of assimilating different theories. So, what has this got to do with a notion that tries to reconcile diversity and directionality with strong unity? What is the role of identity based reduction in the much broader context of the reduction debate in the philosophy of science?

Reduction in the philosophy of science is usually regarded as a relation instantiated by pairs of theories, or scientific models, rather than by "pairs" of properties, event-types or states of affairs. Recall the distinction between *theory* reduction and *ontological* reduction introduced in Sect. 2.1.7:

(Theory Reduction) The predicate '_ reduces to _' expresses a relation of theory reduction only if its arguments in true sentences refer to things such as theories, or scientific models, or fragments of theories.

(Ontological Reduction) The predicate '_ reduces to _' expresses a relation of ontological reduction only if its arguments in true sentences refer to things such as kinds, types, properties, events, substances, or individuals.

The concept of identity-based reduction, which underlies the puzzle of reduction discussed in Chap. 3, can partly be characterized as follows: *a reduces to b only if* $a = b$. Now, obviously, if a theory t reduces to a theory t^*, then it is not the case that $t = t^*$. The so explicated notion of theory-reduction does not, all by itself, give an idea of how identity-based reduction may enter a model of theory reduction. But in fact, identity-based reduction comes in two forms, as a theory-variant as well as an ontological variant. For the ontological version of identity-based reduction, we get the following principle:

(Principle Identity-Based Ontological-Reduction): If a reduces to b, then $a = b$.

This concept of reduction amounts to unity by identification. In order to understand how a concept of theory-reduction may fit the characterization that *the concept of reduction reconciles diversity and directionality with strong unity, without relying on elimination*, consider Ernest Nagel's famous characterization of the reduction-relation:

> [A] reduction is effected when the experimental laws of the secondary science (and if it has an adequate theory, its theory as well) are shown to be the logical consequences of the theoretical assumptions (inclusive of the coordinating definitions) of the primary science. (Nagel 1961, 352)

The basic idea is simple: A theory T_R reduces to a theory T_B iff T_R is derivable from T_B plus the relevant bridge laws (here labeled 'coordinating definitions'), if any, with the contention that often, theory reduction is carried out by reduction of the theory's laws. If we add the remarks Nagel opened his discussion with – namely, that reduction has to be understood as a certain kind of explanation (Nagel 1961, 338) – the core idea of the Nagel model is fully characterized. Reduction is (i) a kind of explanation relation, which (ii) holds between two theories iff (iii) one of these theories is derivable from the other, maybe under specific conditions, with (iv) the help of bridge-laws. Building on this sketch, we can easily illustrate the *identity-case of theory reduction* for the Nagelian model: bridge-principles might be regarded as stating identities. In this case, Nagelian reduction amounts to the following conditional: If T_R reduces to T_B in the identity-sense, then for any kind x of T_R, there is a kind y of T_B, such that $x = y$. A more general variant of the theory-case of identity based reduction can then be identified as follows:

(Principle Identity-Based Theory-Reduction): If T_R reduces to T_B, then the (relevant part of) the ontology of T_R belongs to the ontology of T_B.

The so characterized identity-case of theory-reduction plays a fundamental role for the reduction-debate in the philosophy of science, although not all characterizations of the concept of reduction offered in the philosophy of science coincide with this notion. Sometimes, it will be relatively difficult to uncover the connection between a given proposal and the identity-case of theory reduction: Let the relation defined by a given model of theory-reduction proposed in the literature be $\mathrm{R_{MODEL}}$. Let the identity-case of theory-reduction be $\mathrm{R_{IDENTITY}}$. For the models of theory reduction discussed below, it will be shown that possibly, $x\ \mathrm{R_{MODEL}}\ y$ & $x\ \mathrm{R_{IDENTITY}}\ y$. Thus, it will be shown that at least, these models are compatible with the identity case; often, they explicitly allow for the identity-case as a special, or, as it is sometimes put, as a *limiting* case. Secondly, it will be argued that for any monistic world, any case of $\mathrm{R_{MODEL}}$ will be a case of $\mathrm{R_{IDENTITY}}$, and that, therefore, an understanding of theory-reduction as corresponding to our slogan is tied to an understanding of the identity-case of theory-reduction (and, hence, for an understanding of reduction that suits the monist, we have to understand the identity-case of reduction). Thirdly, it will be argued that as soon as correction enters definitions of reduction, an appropriate understanding of $\mathrm{R_{IDENTITY}}$ is required for understanding $\mathrm{R_{MODEL}}$. Finally, in the next Chapter, the relation between the theory-variant and the ontological variant, that is: the explication proposed in Chaps. 4 and 5, will be addressed, and it will be argued that the latter is *prior* to the former – a view that is not particularly popular among philosophers of science.

In order to assess the role the identity-case of theory reduction plays in the reduction debate in the philosophy of science, I will guide us through the most important (families of) definitions offered in the past 60 years. To keep things simple, I will introduce *canonical re-descriptions* of each of these models, highlighting three different aspects. Recall the sketch of Nagel's model. According to Nagel, reduction is *explanation* by way of *derivation* of one *theory* by another, with the help of *bridge-laws*. This rough characterization comprises three relevant aspects, according to which the different models will be characterized and compared, such that their relation to the identity-case can be evaluated. First, competing characterizations may differ in how they define the relation itself; they may differ in what they substitute for '*R*' in a definition of the following kind: x reduces to $y \leftrightarrow_{\text{def.}} x\ R\ y$. For Nagel, it is '_is derivable from _ with the help of bridge-laws (under specific conditions)'. Secondly, different characterizations may differ in how they conceive of the entities that primarily instantiate the reduction relation. For

Nagel, these are theories. Finally, different characterizations often differ in how they conceive of the relation between *reduction* and *explanation*. For Nagel, reduction is a way to instantiate the (appropriate version of the) DN-model of explanation, and, hence, an explanation relation. The canonical re-descriptions will thus consist in answers to the following three questions:

For a given model of reduction,

(i) How is the reduction relation characterized?
(ii) What are the (primary) candidate-relata for the reduction relation?
(iii) What is the suggested link to explanation?

Having answered these questions for each of the relevant characterizations, we will be in a position to relate them to the identity-case of theory reduction. In the following four sections I will summarize two versions of Nagel's model (7.2), its structuralist counterpart (7.3), Schaffner's extension of this model (7.4) and the characterization pertinent in what Bickle coined 'New Wave' reductionism (7.5). Note that given the rich debate on *epistemological* issues of reduction in the philosophy of science – on how reductions are justified, achieved, constructed, discovered, or how they are connected to scientific methods – a characterization of these models merely in terms of the reductive link, the candidate-relata and the link to explanation may seem rather meager. However, this is all we need in order to properly address the guiding conceptual question that is at issue here: How does identity based theory reduction, as partly characterized by *(Principle Identity-Based Theory-Reduction)*, relate to the most influential characterizations of reduction proposed in the philosophy of science?

7.2 Nagel's Model

The version of Nagel's model introduced above is the "official" version – the one discussed in the literature as *the* Nagel model of reduction, according to which reduction is an explanation-relation, consisting in the derivation of one theory from another theory with, sometimes, the help of bridge-laws. This version covers two cases of reduction, depending on whether bridge-laws are required or not: *non-homogeneous* cases and *homogeneous* cases. Nagel conceives of sciences or theories as developing entities, which undergo changes, even though the vocabulary remains unchanged (but is, presumably, often extended). These successive stages of theories are covered by the notion of *homogeneous* reductions – deduction of an early stage from a later stage of a theory is possible without bridge-laws. *Non-homogeneous* cases of reduction are instantiated by pairs of different theories, employing different vocabularies. Whereas the former variant of reduction did not attract much attention (by Nagel and others), the latter has been intensively discussed ever since Nagel introduced his model in "The Meaning of Reduction in the Natural Science" (Nagel 1949).[1] Based on more elaborate versions of the rough outline just given, the Nagel-model became subject to a number of well-known criticisms worth mentioning:

It is too narrow because it allows for theory reduction only (Sarkar 1992; Darden and Maull 1977, 43; Hull 1974; Wimsatt 1974), whereas an appropriate model would cover cases of reduction of mere models and the like – sciences like biology and neuroscience should be regarded as being possible candidates for reduction, although they do not contain full-fledged theories. The model exemplifies all the shortcomings of the orthodox view on science. For example, it conceives of theories as syntactic entities and of explanation to be cashed out in terms of the DN-model (Hempel and Oppenheim 1948), which has been criticized for a number of reasons, amongst which issues regarding the asymmetry of explanation play a prominent role (for an overview, which mainly focuses on problems arising from reduction as explanation, see Craver 2007, chapter 2, for problems concerning the DN model, see Salmon 1989). The model describes reduction in terms of direct theory-explanation, whereas an appropriate model of reduction should shape the notion in terms of indirect theory-explanation, that is: in terms of explanation of *the phenomena* of a theory (Schaffner 1967, 1993, 423; Kemeny and Oppenheim 1956; Friedman 1982). In addition, there are at least three formal worries worth mentioning: If reduction is derivation plus (sometimes) bridge-laws, then every theory would reduce to itself (because every theory is derivable from itself); moreover, every theory would reduce to every inconsistent theory; and contrary to what one might expect, reduction is not directional – derivability allows for mutual derivability (and there is nothing in the notion of a bridge-law that would rule out this possibility). So, the Nagel-model seems to be in a very bad shape.

Despite these alleged problems, Ernest Nagel's model of theory-reduction shaped the discussion on reduction in the philosophy of science and it received considerable attention in the philosophy of mind (see, for example, Kim (1993, 150 & 248) and Fodor (1981, 150)). More recent approaches to reduction depart from the Nagel model (Dizadji-Bahmani et al. 2010; Bickle 1998, 2003; Schaffner 1993; Churchland 1985; Hooker 1981), and it has been argued that most of these approaches merely echo the Nagel-model instead of proposing fundamentally new interpretations (Endicott 1998, 2001; Dizadji-Bahmani et al. 2010). Moreover, a careful reading of Nagel suggests that what is usually referred to as 'the Nagel-model' is not the model Nagel actually had in mind, but rather an utterly

[1]Let me briefly comment on Nagel's four main publications that are directly concerned with reduction. Nagel (1935) gives an intuitive sketch of reduction that is almost fully lost in his later writings, It is, however, interesting for a number of interpretative purposes discussed in (van Riel 2011); it sheds light onto the relation between reductive and mechanistic explanation. Nagel (1949) is almost fully included in chapter 11 of his 1961 book. The point of departure for his interpreters is chapter 11 of Nagel (1961), so I will mainly focus on this chapter. Nagel (1970) is interesting in three respects: first, it includes an interpretation of bridge laws that is much more precise than the interpretation offered in Nagel (1961); second, it includes a discussion of the role of correction in reduction; and, finally, it sheds light onto issues regarding the question of the relata of the reduction relation (for an interpretation, see my van Riel 2011).

deficient caricature of the *real* Nagel-model (as I argued in van Riel 2011). The present chapter does not deal with Nagel-exegesis; thus, I will confine myself to presenting both, the official as well as the stronger reading, as two candidate-characterizations of reduction.

Given the outline of Nagel's model presented above, questions (i)–(iii) can easily be answered, thereby giving a canonical re-description of Nagel's model:

(The official Nagel model)

(i) *How is the reduction relation characterized?*
The relation is derivation plus bridge-principles. Derivation is to be conceived of as syntactic derivation. Bridge-principles are usually conceived of as universally quantified bi-conditionals, or sometimes as conditionals (1961, 355, fn.5), whose predicates are predicates of the reducing as well as of the reduced theories' laws.

(ii) *What are the (primary) candidate-relata for the reduction relation?*
The primary *relata* of the derivation are *laws* or entire *theories*. Since derivation is a syntactical matter, these should be construed syntactically.

(iii) *What is the suggested link to explanation?*
Officially, instantiating the reduction relation is instantiating an explanation relation. Obviously, the reduction relation is a special case of the relation defined by Hempel and Oppenheim (1948) – a candidate for a characterization of explanation. In this sense, the reduced theory is supposed to be explained by the reducing theory.

So, how does identity-based reduction enter the game? For some cases it may turn out that the predicates connected in a bridge-law are not merely co-extensional (what is required for the bridge-law to be true), but that they also *signify the same kind*. For example, the bridge-law '$\forall x$ x is water iff x is H_2O' expresses a truth if water = H_2O. Moreover, if water = H_2O, then this bridge-law will be true *because* water = H_2O. Hence, the truth of bridge-laws may depend upon the truth of identity statements about the kinds signified by the predicates of the bridge-law. In this sense, Nagel's model covers the case of kind-identity and, correspondingly, identity-based reduction as a limiting case. Moreover, for monistic worlds, the kinds employed by the reduced theory will be identical to kinds of the reducing theory. Then, for any reduced theory, for any reducing theory in a monistic world, the former will reduce to the latter according to Nagel's definition only if it reduces according to a variant of this model in which bridge-laws state identities. If we live in a monistic world and if the official Nagel-model characterizes theory-reduction appropriately, then bridge-laws will be true only if (and if so, in virtue of the fact that) the corresponding identity statements about the kinds signified by the predicates connected in the bridge-laws are true. This generalizes: For monistic worlds, Nagel's official model specifies cases that also instantiate the identity-case of theory reduction. Then, to understand what backs up reduction in these worlds, and, hence, what reductionist monism consists in, we need to understand the identity-case of theory reduction.

An interesting application of Nagelian reduction which reveals its similarity to conceptions of reduction in the philosophy of mind, namely, to functional reduction and to type identity theories in general, has been developed by Esfeld and Sachse (2007, 2011), building on the idea that functional sub-types of high-level kinds can be identified with lower-level kinds. Within this picture the idea of kind- or property-reduction can easily be tied to the idea of reduction of theories or sciences: It is contextualized versions of higher-level theories or sciences which reduce.[2] Esfeld and Sachse describe how this contextualization can take place *within* the high-level science, namely, by constructing fine-grained versions of the high-level types, which can then be linked to the relevant low-level kinds. On this view, the *relata* of reduction are theories and specific bridge-laws.[3] The connection between type-identity theory in the philosophy of mind and theory-reduction, as classically construed, is, thus, obvious: functional reduction plus some version of type-identity theory specifies a way to arrive at bridge-laws within a Nagelian model of reduction.

As already suggested, there are serious doubts about whether or not the classical Nagelian model is an interesting candidate for a characterization of what Nagel had in mind when introducing his explication of reduction. An alternative formulation is this (*cf.* van Riel 2011, for a detailed discussion): It is not the case that Nagel allowed for reduction of full-fledged theories only. Other representational devices, such as models and fragments of theories, could possibly be reduced as well. Reduction, according to Nagel, is a relation holding between a great variety of scientific representational devices, amongst which theories play an important *epistemological* role – for theories, we can easily define reduction and identify particular instances of this relation (Nagel 1961, 345). For models or theory-fragments, it is, according to Nagel, hard to adequately define a precise notion of reduction. Moreover, Nagel drew a sharp distinction between *mere* reductions and *interesting* reductions. Interesting reductions are explanations that consist in deductions, which are carried out with the help of bridge-laws, and they have to meet a number of additional non-formal conditions, such as unification and appropriateness of the reducing theory and the bridge-laws (Nagel 1949, 304 ff.; 1961, 358 ff.). Moreover, if required, correction can be involved in reductions (ibid.).[4] In addition, according to Nagel, bridge-laws are to be regarded as *stating* ontological links (identities or relations among extensions) *a posteriori* (1961, 340 & 354, 1970, 126 ff.). Reduction is not merely the explanation of one theory by another theory; rather, it goes together with explanations of the phenomena of the reduced theory by the reducing theory (what has been labeled 'indirect'

[2] Alternatively, one might suggest that non-contextualized theories reduce in virtue of some relation between their reduction base on the one hand, and its contextualized versions on the other.

[3] The Esfeld-Sachse approach seems to be capable of doing more: It shows the value of the concepts of high-level types in the construction of the relevant contextualized kinds – here dubbed 'sub-types'. (Esfeld and Sachse 2007; Sachse 2007).

[4] This last point is astonishing, since the idea that Nagel's model does not allow for reduction has been the basis for the most prominent criticisms, initially raised by Feyerabend (1962, 1966). However, already Putnam (1965, 206 ff., esp. n. 3) was well aware of the fact that correction could easily be incorporated in Nagel's model of reduction (for similar reasons, Nickles (1973) suggests that these criticisms do not succeed).

theory-reduction, as opposed to 'direct' theory-reduction (this is explicit in the discussion of an example in (Nagel 1961, 366 – see Sect. 4.4.3 for a discussion), and, similarly, in (Nagel 1961, 434)). By the transitivity of explanation, if the fact that *r* explanatorily depends on the fact that *q*, and if the fact that *q* explanatorily depends on the fact that *p*, then the fact that *r* explanatorily depends on the fact that *p*. If you substitute a truth about a phenomenon of the reduced theory for '*r*', and the corresponding truth of the reduced theory for '*q*' (what expresses the explanans for the explanandum that *r* within the reduced theory), and for '*p*' the reduced theory's truth(s) from which '*q*' is derived with the help of bridge-laws, you arrive at the explanation '*r* because *p*'. Hence, theory-explanation gives rise to phenomena explanation. Finally, and not surprisingly, given the ontological backup bridge-laws provide, Nagelian reduction is not a purely epistemological issue[5]; rather, it comes with strong ontological commitments – bridge-laws state ontological links, such as identity-links or nomological connections between extensions (cf. van Riel 2011, 3.3). If this is correct, then we get the following picture of Nagelian reduction (what is here referred to as the "real" Nagel-model):

(The real Nagel model)

(i) *How is the reduction relation characterized?*
The relation is derivation plus bridge-principles. Derivation is to be conceived of as syntactic derivation. Bridge-principles are conceived of as stating identity-links a posteriori, and sometimes as stating a relation among extensions (probably in modal terms, building on nomological necessity or nomological equivalence). Interesting reductions will meet further criteria, such as unification or correction.

(ii) *What are the (primary) candidate-relata for the reduction relation?*
In an idealized version, the primary *relata* of the derivation are sets of *laws* or entire *theories*. However, Nagel was more liberal concerning possible *relata* of the reduction relation – fragments of theories and models seem to be serious candidates for reduction as well.[6]

(iii) *What is the suggested link to explanation?*
Officially, instantiating the reduction relation is instantiating an explanation relation: Obviously, the reduction relation is a special case of the relation defined by Hempel and Oppenheim (1948) – a candidate for a characterization of explanation. However, this should not give the impression that merely the reduced theory is supposed to be explained by the reducing theory. Rather, the reducing theory is supposed to explain the phenomena captured by the reduced theory.

[5]Silberstein (2002, 80 f.) describes Nagelian reduction as epistemological because it is an instance of theory-reduction. Others, like Sarkar (1992, 171) and Hoyningen-Huene (1989, 30), argue that it is epistemological because it is described as an explanation relation.

[6]If this characterization is correct, Nagel anticipates Schaffner's partial reductions (Schaffner 2006).

We will discuss the case of reductions involving correction later (Sects. 7.4 and 7.5) – it will turn out that again, an appropriate understanding of the identity-case is a prerequisite for understanding models of reduction that build upon correction of the reduced theory by the reducing theory; let us, for the moment, focus on the limiting case in which bridge-laws, in the real Nagel-model, state identities, i.e. have the form '$a = b$' (or are true in virtue of a truth of this form). Here, obviously, identity-based reduction is crucial for our understanding of one of the most influential models of reduction in the philosophy of science. To account for reduction relations within a monistic world, the identity-case of Nagelian reduction suggests itself; again, understanding reduction-relations in such worlds requires us to understand the identity-case of theory-reduction. As we will see, other models of reduction in the philosophy of science are similar to the Nagel-model in this respect. Before discussing Schaffner's *General Reduction Replacement Model* and models of *New Wave* reduction, in which the aspect of correction figures prominently, let us turn to the structuralist counterpart of the Nagel-model.

7.3 Structuralism

Structuralist approaches to reduction differ from the Nagelian model primarily in that they interpret the *relata* of the reduction relation in a fundamentally different way. To fully understand the structuralist discussion of reduction and its relation to discussions of reduction in other parts of the philosophy of science, it is worth reflecting briefly upon the structuralist conception of a theory (if you are familiar with this conception, you may want to skip the next paragraph).

The structuralist conception of theories arose as a response to what was regarded as a fundamental problem for the empiricist picture of science. For example, the notion of a *theoretical term* (as, for example, described in (Carnap 1958)) was regarded as highly problematic (Sneed 1971). Moreover, part of the classical structuralist program consists in the idea that the philosophy of science, properly construed, pursues an ideal of *rational reconstruction* of science (cf. Stegmüller 1979; Balzer 1984, 331), which is to be understood as a reconstruction of actual scientific developments, such that *extension first*-criteria are, so to speak, built into the program from the beginning. For example, the idea of a *non-extensionally defined set (or a family of sets) of intended applications* was introduced to handle the behavior of scientists when they were confronted with an alleged failure of a theory – being an intended application gives a criterion to judge whether or not a failed application is a failure of the theory or a failure in applying the theory on a given occasion. According to the structuralist picture (Sneed 1971; further developed in: Stegmüller 1979), to give an account of a theory T is to define the predicate '_is a (model of) T' in set-theoretical terms. A theory basically consists of a set M, a set of potential models M_P, a set of potential partial models M_{PP}, and a set of intended applications *I*.[7] The set M is the set of actual models, that is: complexes of objects that fulfill the theory's laws. M_P is the class of potential models – intuitively, the class of objects that can be described using the resources given by M *without* being such that they necessarily obey the laws given for M. Thus, M is a subclass of M_P. The definition of the class M_{PP} is based on the notion of a theoretical term as it occurs in the structuralist picture: here, terms are not theoretical *simpliciter*, but rather theoretical with respect to some theory (depending on the position they occupy within this theory). M_{PP}, then, is the class of models that are such that they do not comprise the theoretical elements (intuitively speaking: the properties or aspects referred to or picked out by the theoretical terms). In addition, a class of constraints has been added, C, which connects different models of a theory, say, by fixing that for some theory identity of objects goes together with identity of some specific property (at a time). These aspects form the theoretical core of a theory: K. Finally, we get the set of intended applications, *I*, which is supposed to be "open", or non-extensionally individuated, or (as I would like to put it in order to avoid misuse of the well defined notion of a set) a family of related sets of empirical objects we intend to describe using the theory in question. The relation between the different variants or stages of *I* is, as Stegmüller suggests, best thought of as being pragmatically fixed (Stegmüller 1979). Thus, a theory is (at least) a pair <K,I>. It is easy to see that the conception of a theory is, according to this picture, different from the picture of the so-called 'received view' according to which theories are sets of statements.

In accordance with the postulated ideal of a rational reconstruction of scientific developments, structuralists often focused on cases of mere replacement, or on similarity-relations between theories, which are cashed out in terms of approximation or (semantic counterparts of) derivation under ideal conditions (Moulines 1984; Scheibe 1999; Stegmüller 1979, 1986, ch. 4). Consequently, these conceptions do not straightforwardly translate into definitions of the concept at issue here. At the same time, structuralists tried to capture interesting possible cases of reduction, such as reduction of psychology to physiology. The first influential attempt to characterize this sort of reduction in semantic terms was made by Patrick Suppes. He basically replaced the notion of syntactic derivation by a notion of (partial) structural equivalence, which, in turn, is characterized via the notion of an isomorphism:

> Many of the problems formulated in connection with the question of reducing one science to another may be formulated as a series of problems using the notion of a representation theorem for the models of a theory. For instance, the thesis that psychology may be reduced to physiology would be for many people appropriately established if one could show that, for any model of a psychological theory, it was possible to construct an isomorphic model within physiological theory. (Suppes 1967, 59)

[7]The characterization given here is based on Balzer and Moulines (1996, 12–13). The authors are concerned with larger structures of theory-nets and theory-*holons*. Some aspects they include in the notion of a theory seem to transcend the minimal requirement. I merely focus on the minimal aspects, because they suffice to give the picture needed for present purposes.

If we take this to be an appropriate informal characterization of the reduction relation, reduction amounts to (partial) structural equivalence between the reduced theory and (a part of) the reducing theory. As Schaffner has pointed out, this is too weak a notion to be useful, and (official) Nagelian reduction is a special case of Suppes' model. That is: Whenever there is a Nagel-reduction, then there is a Suppes-reduction. Moreover, Suppes' model covers cases that should not be covered (Schaffner 1967). The strategy of cashing out reduction in terms of an *isomorphism* or *analogy-relation* has been pursued by Sneed (1971) and Stegmüller (1979). The basic problem of these accounts has been pointed out by Moulines (1984, 55; a similar idea can be found in Hoering 1984). Moulines' insight can be regarded as a specification of what gave rise to Schaffner's objection, namely, that arbitrary reductions are not ruled out as long as we try to cash out reduction in purely structural terms:

> There is at least one further aspect of reduction that is overlooked by [the relevant scheme, RvR]. This is what I would like t call "the ontological aspect". I wish to argue that, for a complete picture of a reductive relationship between two theories, one has to take into account some sort of relation between the respective domains. Otherwise, when confronted with a particular example of a reductive pair, we would feel that all we have is an *ad hoc* mathematical relationship between two sets of structures, perhaps by chance having the mathematical properties we require of reduction but not really telling something about "the world". We could have a reductive relationship between two theories that are completely alien to each other. (Moulines 1984, 55)

This sort of criticism amounts to the fact that definitions that do not respect the ontological dimension are too weak – the same criticism Schaffner put forward.[8] Secondly, it is a worry similar to the one that inspired Nagel to come up with additional, non-formal conditions that enable us to distinguish between interesting and non-interesting cases of reduction (see Sect. 7.2). However, there is an important difference between Nagel's worry concerning his model and the worry about structuralist conceptions: the former model already includes reference to the ontology because bridge-laws are cashed out in terms of ontology or at least imply ontological truths about the domains and kinds of any pair of reduced and reducing theories. Structuralist models do not – structural equivalence, independent of how this notion is cashed out in detail, neither implies the existence of intersections of the respective domains, nor the identity of elements of the sets that form the relevant model structure. In this context, it is worth mentioning another problem. Nagel's official model has frequently been criticized because it focuses on candidate *relata* that have to fulfill rather strong criteria; they must be full-fledged theories.

[8] Moulines himself tried to face the problem introducing further purely formal conditions on reduction which are supposed to exclude problematic cases (Moulines 1984). This idea was picked up by Bickle (1998, 79ff.). Both, however, admit that these conditions, which are, again, specified in purely formal terms, will not rule out all possible counterexamples (Bickle 1998, 79; Moulines 1984, 55). It should be clear that, independent of how we shape the formal additional criteria, we will not end up with better candidates than those given by straightforward ontological descriptions that relevantly employ the concept of *identity*.

What I have coined the "real" Nagel-model does not have this feature. Structuralist models, however, are, as it seems, bound to restrict the domain of the reduction-relation to full-fledged theories.[9] Otherwise, notions of structural similarity become problematic. Only if a science or a theory can be reconstructed in structuralist terms, we have a candidate for reduction. But there are serious doubts that sciences such as economy, political science and the like can be reconstructed in this way. Thus, an extension of any structuralist reduction model to the "more pedestrian parts of the scientific enterprise", (as Feyerabend (1962, 28) put it raising his criticism against Nagel), might turn out to be difficult.

Let us finally move to the suggested link between reduction and explanation. In contrast to Nagel, structuralists did not introduce their models of the reduction-relation as models of an explanatory relation. Rather, it is introduced as a relation covering specific kinds of theory-change. Links to explanation have been suggested in passing only. The most dominant view on the link between reduction and explanation is based on an intuitive, non-formal understanding of systematization. Sneed (1971, 218) suggests that the reducing theory can explain, in the sense of systematize, everything the reduced theory can explain.

Let us thus turn to the canonical re-description:

(The structuralist model)

(i) *How is the reduction relation characterized?*
The relation is described in terms of structural equivalence. As pointed out by Moulines, an ontological link is required for the structuralist model to make sense of how the term 'reduction' is usually used.

(ii) *What are the (primary) candidate-relata for the reduction relation?*
The primary relata are theories in the structuralist sense. As secondary relata, one may consider any theory that can be reconstructed in structuralist terms.

(iii) *What is the suggested link to explanation?*
In passing, it has been suggested that reduction goes together with explanation in the sense of systematization: Phenomena of the reduced science are systematized by the reducing science.

The moral to be drawn is this: Depending on how strict we set our standards for comparison, semantic approaches will turn out to be fundamentally new or the same old Nagelian story retold. Adding an appropriate link, structuralist approaches will be similar to the Nagel-model. And depending on the ontological link required to make structuralist versions of reduction capture the intuitive notion of reduction (or, if you like, all possible intended applications of the concept), whether or not identity-based reduction is covered will be obvious. If the link is or amounts to identity among kinds, then, again, for any pair of reduced and reducing theory, the

[9] It is, however, important to see that structuralist notions of a theory do not depend upon the idea that an actual theory has to conform to structuralist standards; rather, the idea is that it can be reconstructed in a structuralist way.

kinds of the former will figure among the kinds of the latter. If weaker ontological links are introduced, which allow for but do not require cross-theoretic identities, then, again, the model will cover such reductions as a limiting case, and it will be relevant to account for cases of reduction in monistic worlds (and, hence, for an understanding of reduction that suits reductionist monism). Let us now turn to a model of reduction that was developed at the same time as structuralists developed their models. It had an important impact on more recent versions of reduction, namely, on versions of New-Wave reduction.

7.4 Schaffner's Reduction-Replacement Model

The *General Reduction Replacement Model* (henceforth: GRR model) proposed by Schaffner in his (1993, chapter 9), is the most recent version of a model he developed in a series of papers before (1967, 1974, 1976), and it reflects part of the criticisms raised against the Nagel-model, some of which have been mentioned above (Sect. 7.2). Most notably, as its title suggests, the *General Reduction Replacement Model* tries to overcome alleged problems that are due to the fact that the classical Nagel-model fails to incorporate theory-successions that comprise an element of correction, which is usually found in actual cases of theory-succession. Thus, Schaffner puts emphasis on the investigation of scientific change or alleged cases of reduction (Schaffner 1993, 515 f), and, consequently, his model relies on the possibility of correction. Nevertheless, the model is designed to cover Nagelian reduction based on identity as a limiting case (Schaffner 1993, 430). Moreover, Schaffner is ready to accept that bridge-laws may state identities (Schaffner 1993, Sect. 7.4.2). His model thus covers the relevant kind of reduction, which goes together with cross-theoretic identities. A careful discussion of the model is in place; Schaffner's model can be used as a paradigmatic example of models of reduction that try to incorporate an aspect of correction, an issue that is, for example, also of particular importance in the context of models of *New-Wave* reduction. Moreover, Schaffner's model can be used to get a more thorough idea of the relation between *extension first* and *criteria first* approaches to reduction, or between attempts to model scientific change and attempts to explicate the concept of reduction.

7.4.1 A Sketch of the Schaffner Model

Kemeny and Oppenheim believed that if we tried to define a model of reduction that covers scientific change including correction then the problem of reduction would become 'hopelessly complex' (Kemeny and Oppenheim 1956, 13). Despite Kemeny's and Oppenheim's pessimistic outlook, those who worked on reduction in the post-Nagelian era tried to fully characterize this aspect. And so did Schaffner, arguing that it would be desirable to have a model that covers "actual", rather than merely ideal cases of reduction (see, for example, Schaffner 1993, 426 and 427 f).[10] His definition or model is rather complex, comprising several disjuncts, which are supposed to guarantee that both, genuine reduction *and* different sorts of replacement are met, from complete replacement to some sort of correction in the reduced theory. The main points can be made building on the following upshot:

Schaffner is explicitly liberal concerning the *relata* of the reduction relation. Models as well as theories can enter the reduction-relation. Basically, Schaffner tries to give a model of reduction and replacement-relations as theory successions which occur in virtue of a combination of, or in virtue of one of, two different sources: cross-theoretic identities and mere replacement. For ideal cases, we get reduction that is roughly Nagelian in spirit, and that is based on cross-theoretic identities. For less ideal (and, thus, most actual) cases, we get reduction by way of correction. '_being corrected by_' (in the relevant sense) is characterized as follows: A theory t_1 is corrected by a theory t_2 iff there is a variant of t_1* (probably not identical to t_1), such that t_1* uses the vocabulary of t_1, and t_1* reduces to t_2 in the more demanding, ideal sense (here, Schaffner employs the term 'strong analogy', which is seemingly taken to be primitive).[11] The case at the opposite side of the spectrum, the case of complete replacement, or elimination, is not interesting for our present concern. Following Nagel to a certain extent, Schaffner suggests that reductions are explanations in one of the following two senses: either, the reducing theory explains the reduced theory and it 'indicates' why it worked as well as it did, or it explains the domain of the reduced theory (Schaffner 1993, 429). This rough sketch suffices to characterize Schaffner's model in terms of a canonical re-description, and to identify the role identity-based reductions play within this model. Before turning to these issues, let me pause for a moment to make a small detour; the following excursus discusses Schaffner's model in more detail, illustrates the difference between *replacement*-issues on the one hand, and *reduction*-issues on the other, and thereby picks up the topic of methodology discussed in Chap. 2.

---

[10]This is reflected by Schaffner when he describes the *method* he employs. It is to be conceived of as 'logical pragmatism' the basic idea of which can be cashed out as follows: Try to incorporate not only the relevant conceptual aspects, but also what actually happens in science; pragmatic or procedural aspects of theorizing, deciding and so forth are relevant, too. These different aspects are to be combined in a model that mirrors the multidimensional endeavor of the pragmatics, epistemics and logic of science (Schaffner 1993, 515 f.).

[11]According to the GRR-model, it is possible to formulate the corrected replaced theory in the language of the replaced theory or in the language of the replacing theory, or, differently: To derive a theory-variant from the replacing theory plus boundary conditions in order to get the mapping we need for reduction/replacement. This is not our target here, so I ignore this aspect.

7.4.1.1 Excursus: Schaffner's Model and the Replacement/Reduction Distinction

Schaffner's reduction-replacement model nicely illustrates that there is a fundamental difference between reduction and replacement, and it suffers from the unfortunate consequences the combination of these two issues may have. Here is Schaffner's model:

T_B – the reducing theory/model
T_B* – the "corrected" reducing theory/model
T_R – the original reduced theory/model
T_R* – the "corrected" reduced theory/model
(In those cases where only one of T_B or T_{B*} is the reducing theory, we shall use the expression $T_{B(*)}$.)

Reduction in the most general sense occurs if and only if:

(1)(a) All primitive terms of T_R* are associated with one or more of the terms of T_B(*), such that:
(i) T_R* (entities) = function (T_B(*) (entities))
(ii) T_R(*) (predicates) = function (T_B(*)(predicates))[...]

or

(1)(b) The domain of T_R* be connectable with T_B(*) via new correspondence rules. (Condition of generalized connectability.)

(2)(a) Given fulfillment of condition (1)(a), that T_R* be derivable from T_B(*) supplemented with (1)(a)(i) and (1)(a)(ii) functions.

or

(2)(b) Given fulfillment of condition (1)(b) the domain of T_R be derivable from T_B(*) supplemented with the new correspondence rules. (Condition of generalized derivability.)

(3) In case (1)(a) and (2)(a) are met, T_R* corrects T_R, that is, T_R* makes more accurate predictions. In case (1)(b) and (2)(b) are met, it may be the case that T_B* makes more accurate predictions in T_R's domain than did T_R.

(4)(a) T_R is explained by T_B(*) in that T_R and T_R* are strongly analogous, and T_B(*) indicates why T_R worked as well as it did historically.

or

(4)(b) T_R's domain is explained by T_B(*) even when T_R is replaced. (Schaffner 1993, 429)

Let me first comment on these conditions. Two terminological remarks are in place: A *correspondence rule* links the observational language to the theoretical language (the details are not needed to judge the model) (Schaffner 1969). A theory's domain is "a complex of empirical results which either are accounted for [...] and/or *should be* accounted for [by the completely developed version of the theory it is the domain of, *RvR*]" (Schaffner 1993, 428).

Now, the disjunctive nature of this model makes it hard to uncover what it actually states. The model is disjunctive because it tries to capture genuine reductions as well as mere replacements as well as mixed cases at once. This leads to serious problems.

First: Schaffner wants strong Nagelian reduction (i.e. reduction without correction) to be covered. This version of reduction is only captured if there are no corrections involved (that is: If there is no replacement). However, if there is no correction of the reduced theory, condition (1)(a)(i) and condition (1)(b) cannot be met. And only if at least one of these conditions is met, (1) is fulfilled. And only if (1) is fulfilled, we have a case of reduction or replacement according to the GRR model. Note that, therefore, we should interpret 'T_R*' as standing for corrected versions of T_R *as well as* for T_R itself (which would dissolve the problem) at least in 1(b).[12] It does not work in (1)(a): Assume that it is ambiguous in (1)(a)(i). This would contradict condition (3) which explicitly states that the relevant difference between T_R and T_R* is given in (1)(a). Assume that it is ambiguous in (1)(b): Then, we have a set of cases defined by (1)(b), (2)(b), (3) and (4) which *covers* examples of straightforward reductions (together with lots of other versions of inter-theory-relations). However, it does not give us the conceptual resources to distinguish between these cases, because many different cases are covered, including, for example, mere replacements in the radical sense, according to which no important connection between the expressions of the two theories *via* a corrected version of the reduced theory is required. One might think that in (1)(b) the '*' is just a typo. On the other hand, it is (1)(a) which covers identity (here, a mapping of the ontology of one theory onto the ontology of another theory). This mapping should be possible in the case of Nagel-reductions. (1)(b), instead, is designed to cover mere replacements (Schaffner 1993, 428). So, it is difficult to see how the Nagel model fits into this picture.

Note secondly that the model is circular, if we assume that 'reduction in its most general sense' is conceptually dependent on the notion of replacement, which occurs in (4)(b).

Note thirdly that the idea of theory-explanation is explicated in an odd way ((4)(a)): What is introduced as explanation *of a theory by another theory* turns out to be explanation of *why a theory worked as well as it did* (this characterization is also pertinent in Sklar (1967) and Churchland (1986, 283)). That is: A sentence predicating something of a theory is explained rather than the theory itself.

Fourthly: the domain of a theory is not derivable, because the domain is not a linguistic or representational item. Rather, *a full description* of the domain would be derivable. Thus, (2)(b) has to be restated. This difference might turn out to be extremely relevant, if the relevant differences between theories are primarily descriptive in nature.

Fifthly: One might wonder whether or not some of these conditions are redundant. (2), for example, seems to follow from (1) (and (2)(a) from (1)(a) and (2)(b) from (1)(b)). If any primitive term of T_R is connected to a term in T_B by a function that maps the referents of (or what is picked out by) the terms of T_R onto referents of (or on what is picked out by expressions) of T_B, then, introducing an expression of that function, we have derivability. The same holds for the appropriate (see point

[12]In an earlier model, the *general reduction paradigm*, Schaffner uses a similar framework and contends that such an ambiguity is possible (Schaffner 1967, 145).

four above) version of (2)(b): We easily derive descriptions of T_R's domain from its reducer and the correspondence rules, if the correspondence rules come in the appropriate form.

Sixthly: If T_R* is defined as being the corrected version of T_R, then the first sentence of (3) is redundant.

Seventh: Incorporating expressions like 'even when' (4)(b) and 'may' (3) makes it hard to uncover what Schaffner had in mind.

According to a charitable reading, the GRR model is not a definition, but rather a summary of interesting features of reductions and replacements, amongst which we maybe find features that make for a definition or an explication. According to an uncharitable reading, based on a literal understanding of 'definition', it is misguided. Most of the problematic features just mentioned could probably have been avoided. But some problems seem to be tied to the idea that in order to "understand" reduction and replacement all at once, we should be sensitive to pragmatic, epistemic, procedural, metaphysical and logical issues of both cases all at once. A case in point is this: Why go disjunctive? Schaffner does so to cover cases of "a continuum of reduction relations" (Schaffner 1993, 428), which arises from different combinations possible in the disjunctive definition.[13] But some conditions incorporated in this model are based on observations of alleged cases of reductions, such as the idea that we should be able to explain why a theory worked as well as it did (historically). Others seem to reflect intuitions in the earlier reduction debate, intuitions concerning criteria relations between theories have to meet in order to count as reduction relations. Thus, the present model does not only shift between reduction and replacement, and, thus, between conceptual, epistemic, and procedural issues. It also refers to what has been called 'extension first' criteria as well as 'criteria first' criteria in Chap. 2. Note that it is not necessarily the case that this has devastating results; but it should be obvious that the justification, the motivation, and the explication of these rather different claims, with different goals and based on different procedures, could have been made more transparent if the different targets and methods had been dealt with separately.

Depending on what one expects from a model of reduction, this model is disadvantageous or not. In the literature, Schaffner's model is often treated as an amendment of the Nagelian model – some philosophers seem to believe that it plays in the same game as the Nagel-model does (see, for example, Dizadji Bahmani et al. 2010). However, it should be regarded as an extension rather than an amendment. What makes for the extension relies on a completely different concept (mere replacement); hence, one might ask whether or not it is a *second* (and more frequent) type of theory-succession which is addressed here, and which is then combined with an independent (mind: conceptually independent) concept of reduction. From a formal standpoint, this is absolutely legitimate. But tactically, this

[13]There are exactly 8 ways to make the right-hand side of the bi-conditional of (3) in the model given above true, if we interpret 'it may be that p' as used in (3) as implying 'p or not p'. Otherwise, there are 6. Therefore, the term 'continuum' is highly misleading.

is, I think, a problematic idea. We have two separate issues, relatively independent of each other. Why not address them separately? Looking at Schaffner's model, one might think that Kemeny and Oppenheim were right: Such definitions become 'hopelessly complex' (Kemeny and Oppenheim 1956, 13).

In the light of this brief discussion, I submit that the informal sketch given above will serve the purpose of this book. Reflection on the problems Schaffner's definition poses shows that, even if these problems could be overcome, it is not clear what is to be gained by discussing and characterizing replacement or a model covering actual paradigmatic cases of reduction on a par with the concept of reduction. Thus, building on the brief version described in the previous section, I will now turn back to the initial question and show that an understanding of the concept of reduction is required for understanding those parts of the model that are supposed to cover cases of replacements involving corrections.

7.4.2 The Role of Reduction and Identity in the Schaffner Model

Based on the characterization suggested above, we arrive at the following description of the *General Reduction Replacement Model*:

(The General Reduction Replacement Model)

(i) *How is the reduction relation characterized?*
Ignoring the case of mere replacement, the relation is described in terms of a mapping function, either directly from terms of the reducing theory to the reduced theory, or from terms of the reducing theory to a corrected version of the reduced theory. This mapping is not further specified; however, it seems that in at least some cases of reduction, a term is mapped onto another just in case of co-designation.[14]

(ii) *What are the (primary) candidate-relata for the reduction relation?*
The primary relata are theories or, more liberally, scientific representational devices, such as model-descriptions.

(iii) *What is the suggested link to explanation?*
According to Schaffner, in the case of reduction, the reducing theory explains why the reduced theory worked as well as it did, or it explains the domain of the reduced theory (i.e. what the reduced theory was supposed to explain).

---

[14]In addition, Schaffner mentions that part-whole relation might hold between the objects of the reduced and the reducing sciences' domains. We shall come back to this later on (Chap. 8). Note that the way Schaffner's model is presented here, all cases of 1a/b and 2a/b combinations in Schaffner's definition are met. Depending on how we interpret "derivability of domain" in Schaffner's model, this condition will be the one that does maybe not require cross-theoretic identities (if, say, an extensional relation is sufficient).

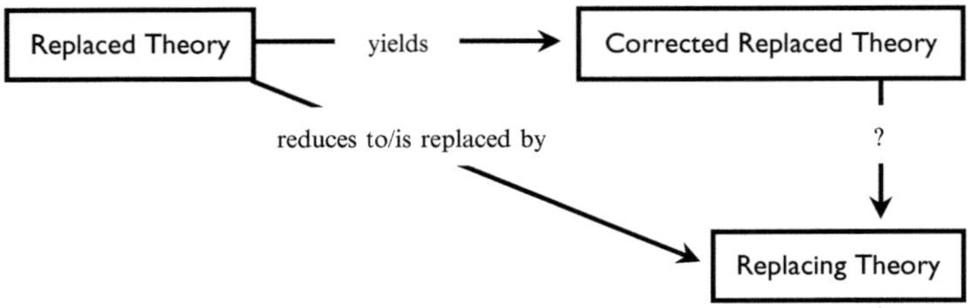

Fig. 7.1 Schaffner reduction

The identity-case of reduction is covered by the model. This is the case of reduction relevant for the monist. But what if correction is involved? Consider the following illustration (Fig. 7.1):

What is the relation between the corrected replaced theory and the reducing theory, the relation marked with a question-mark in Fig. 7.1? In case the relevant fragment of the reducing theory is ontologically equivalent to the corrected replaced theory, we have a case of reduction based on cross-theoretic identities. At least possibly, the corrected theory employs kind-terms that designate the same kinds their counterparts in the replacing or reducing theory designate. Thus, even in the case of replacement, an understanding of how the identity-case of reduction works is required: This is so because the very idea of correction, as cashed out here, builds upon the idea that the identity case may turn out to be responsible for the reduction of the corrected replaced theory to the replacing theory. Thus, on such an interpretation of correction, an understanding even of some forms of *elimination* might require an understanding of the identity-case of reduction. This is so because cases of elimination, which are not cases of outright rejection, will be mediated by pairs of theories that instantiate a relation of identity-based reduction.

It is worth noting in passing that at first sight, correction might be regarded as explaining the *directionality* of theory-reduction. However, we could have started with the corrected theory in the first place, without ever assuming the non-corrected replaced theory, and still, we would have a case of reduction. Reduction of the corrected replaced theory is independent of there being correction. In this sense, the identity-case is relevant even if there is no actual case that works without correction, as long as replacements could possibly be reconstructed as reductions of corrected replaced theories. The possibility that these theories are connectable via bridge-laws that are based on identities is not excluded, and, according to Schaffner, explicitly permitted (see conditions 1(a) and 2(a) in the *Excursus*). This does not contradict our initial diagnosis that replacement does not pose any conceptual problem. It does not. But it also does not help us finding an answer to the question of what *reduction* consists in either.

The relation between *corrected* theories and their reducing counterparts should be regarded as a relation that, all at once, reconciles diversity and directionality with strong unity, without relying on elimination. At the same time, the model covers cases that do rely on elimination; this is the idea of combining characterizations of replacement and reduction. As we shall see in the next section, the same holds for models put forward in the field of New-Wave reduction, which mainly differ from the Nagelian and from Schaffner's model in their characterizations of the epistemological links between the *relata* of the reduction relation – a topic that will be largely ignored here, since it does not contribute to the present chapter's aim of locating the identity case of reduction within the various models of reduction proposed in the philosophy of science.

7.5 New Wave Reductionism

New Wave reductionists take it that their models give an appropriate reconstruction of a specific kind of scientific development (Churchland 1986, 279 ff.; Bickle 1998, chapter 1). At the same time, however, this movement is associated with a *reductionist metaphysical position*: Arguing that neuroscience reduces or eliminates psychology, some protagonists try to show that we should embrace reductionism resp. eliminativism concerning the mind (for example, Paul Churchland 1981, 1985; Patricia Churchland 1986; Bickle 1998, 2003). Thus, there is a tendency to continue debates originating from the philosophy of mind tradition. The models developed by Paul Churchland, Clifford Hooker (1981), and John Bickle are clearly inspired by Nagelian and semanticist models (that is: by the debate in the philosophy of science) and can be interpreted independent of how they function in arguments for reductionism concerning the mind. I will focus on a small fragment of what New-Wave reductionism as a research program is concerned with. This is, again, motivated by the fact that we merely need to locate the case of identity based reduction within this framework, and to re-describe the core idea in the form of our canonical re-description. In these respects, models of *New Wave Reductionism* bear striking similarities to Schaffner's model (as has been pointed out, for example, by Schaffner himself (2012)).

Models of reduction that are nowadays commonly subsumed under the heading of *New Wave Reductionism* have been developed in a series of articles and books by Clifford Hooker (1981), Paul Churchland (1979, 1985) and, more recently, by John Bickle (esp. 1998), and they build on some aspects of Schaffner's model (1993) and earlier versions of it (Schaffner 1967, 1976). Here is a formulation of Hooker's definition, which nicely captures the basic strategy:

> Within T_B construct an analog, T^*_R, of T_R under certain conditions C_R such that T_B and C_R entail T^*_R and argue that the analog relation, AR, between T^*_R and T_R warrants claiming (some kind of) reduction relation, R, between T_R and T_B. Thus (T_B & $C_R \rightarrow T^*_R$) and (T^*_R AR T_R) warrants (T_B R T_R). (Hooker 1981, 49)[15]

The conditions, CR, consist of limiting assumptions and boundary conditions that guarantee that if T_B is more comprehensive than T_R, the application of elements

[15]This is equivalent to Paul Churchland's (1985) and Bickle's (1992) model of reduction.

of T_B's vocabulary is restricted to the domain relevant for T_R. The idea is to effect reductions that include corrections by working on the reducing theory (here 'T_B'), rather than by working with corrections formulated in the vocabulary of the reduced theory. Ronald Endicott nicely sums up the relevant features of New Wave Reductionism:

> (i) New-wave construction: the basic reducing T_B, not the original reduced T_R, supplies the conceptual resources for constructing the corrected T_R*.
> (ii) New-wave deduction: the corrected T_R*, not the original reduced T_R, is deduced from the basic reducing T_B.
> (iii) New-wave relation: there is a required analogical relation, not bridge laws, between the reduced T_R and the corrected T_R*.
> (iv) New-wave continuum: there is a continuum of strong to weak analogies between the reduced T_R and the corrected T_R*, with the strong relations justifying retention and the weak relations justifying replacement of the ontology of T_R. (Endicott 1998, 56)

Thus, the general framework is in accordance with Schaffner's model: Models of reduction should be able to cover cases of replacements based on correction within the base theory. Like the Schaffner-model, it is supposed to cover a continuum of relations between theories that make for reduction or replacement. Within certain boundary conditions (under which T_R* is derived), the reducing theory reflects aspects of the reduced theory. Reflection of aspects comes in degrees: The analogical relation mentioned in (iv) may come in various ways, some of which make for straightforward reductions whilst others do not. Identity (a limiting case for the continuum), for example, ensures that the ontology of the reduced theory is preserved by the reducing theory.

So, what is the basic refinement of New Wave reduction? According to Bickle (1998, 29, 1992, 224) and Paul Churchland (1985, 11), avoiding reference to bridge-principles has a great *epistemological* advantage: The epistemological virtue of reduction does not depend upon knowledge of bridge-principles. Rather, *comparison of reduced and reducing theories justifies identity claims.*[16] How reductions are carried out, and how we arrive at knowledge about ontological links that constitute Nagelian bridge-principles, is not our target, and it should be clear that it is not the target of any attempt to define the reduction relation. Rather, this is an epistemological or a pragmatic aspect of reductionist endeavors. And even Nagel seems to be more liberal than one might think at first sight: He allows for bridge-principles to be empirical hypotheses. These hypotheses are based on comparison between different theories, rather than being prior to any comparison (Nagel 1961, 354). Accordingly, it has been argued that New Wave reduction collapses into

[16] A similar point was made by Kim. Kim seems to oppose the Nagelian picture, stating that it uses bridge-laws as premises without explaining them, and, thus, fails to give a model of explanatory reduction (Kim 1998, 90). In addition, he describes bridge-laws as stating contingent facts, and, therefore, not identities. As I have argued in (van Riel 2011) the term 'contingent' in Nagel's sense is an epistemological term. The epistemological or psychological role of bridge-laws is not of interest to us, and it is this role Kim focuses on. Marras (2005, 351) and Block and Stalnaker (1999, 28) have pointed out that bridge-principles acquire their epistemological status *after* the reduction has been carried out (this is also the point of New Wave reductionism).

Nagelian reduction (more or less broadly construed, for example taking Schaffner's version to be a Nagelian model (Endicott 1998, 2001; Dizadji-Bahmani et al 2010)).

The job done by bridge-laws within Nagelian models, or the job done by structural relations within structuralist models of reduction is here supposed to be done by a correlation between expressions of the reduced and the adjusted reducing theory. New-Wave reductionists typically allow for a number of relations backing up these correlations of expressions. Following Bickle, we can conceive of them as sets of *ordered pairs* of elements of the descriptive parts of the vocabularies of the two theories, which enable us to judge the degree of similarity between the derived image (which is formulated in the language of the reducing theory) and the reduced theory (Bickle 1992, 223). Reduction is associated with a space of theory-relations ranging from "perfectly smooth" or "retentive" reductions to "bumpy" reductions, which are best understood as mere replacements (Bickle 1992, 223; Hooker 1981, 45). For our present concern, the former case is more interesting. Bickle describes it as follows: (I_N is the derived image and T_O is the reduced or the "old" theory):

> In cases lying at or near the retentive endpoint, the I_N is the exactly equipotent isomorphic image of the T_O, and no counterfactual limiting assumptions or boundary conditions are required for the derivation of I_N. (Bickle 1992, 223)

In this case, the pairing of the terms corresponds to identity of referents (Bickle 1992, 224) and I_N can be directly obtained from T_R (the reducing theory), whereas in cases on the opposite point of the spectrum, pairing is achieved only by reference to counterfactual limiting assumptions and boundary conditions. In this case, ontology is eliminated, like in the case of reduction of phlogiston-theory. Note, however, that any similarity or analogical relation should be given a reading that allows for it to come in degrees, such that there is a spectrum of reductions/replacements.[17] The ontological links are, then, identity (smooth reduction and identification) or non-identity, the latter coming in degrees. In the worst case, we have bumpy reduction including total elimination of the ontology of the reduced theory.

In accordance with the tradition, the *relata* of the reduction relation are taken to be theories (Hooker 1981; Patricia Churchland 1986; 278; Bickle 1998, 2003). Churchland explicitly states that theory-reduction is prior to ontological reduction:

> Statements that a phenomenon P_R reduces to another phenomenon P_B are derivative upon the more basic claim that the theory that characterizes the first reduces to the theory that characterizes the second. (Churchland 1986, 278)

This idea is tied to the hope that we can avoid ontological confusion by rephrasing "ontological reduction" talk in terms of theory-reduction (Patricia Churchland 1986, 279).[18] The notion of a theory is, however, more liberal than the classical one,

---

[17]For a detailed introduction, see chapter 1 and 2 of Bickle's (1998).

[18]Bickle (1998) gave a semantic interpretation of New-Wave reduction which is not listed here. As it stands, it seems to inherit problems of both, semantic approaches and classical ways of framing New-Wave reduction. Bickle seems to suppose that the semantic interpretation matches the intuitive one which forms our target.

such that we do not relevantly alter the model when describing it as allowing for models and fragments of theories as relata of the reduction relation.

New-Wave reductionism repeats and adds a number of non-formal aspects to the list of additional criteria we are already familiar with. Let me add five of them, all of which state that there is a specific link between reduction and explanation:

(i) Reductions guarantee explanatory unification (Patricia Churchland 1986, 279, Hooker 1981).
(ii) Reductions yield ontological unification (Patricia Churchland 1986, 280, Hooker 1981).
(iii) In virtue of some reductions, the range of explanation is expanded (Patricia Churchland 1986, 283).
(iv) The old theory is explained by the reducing theory (Patricia Churchland 1986, 283).
(v) The new theory explains why the old one worked as well as it did (Patricia Churchland 1986, 283).

Thus, we are now in a position to give the canonical description of the model:

(New Wave Reduction)

(i) *How is the reduction relation characterized?*
The reduction relation is described as a mapping function that takes terms as arguments and maps these onto terms. The mapping function may correspond to different semantic relations between the pairs of terms it yields: from co-intensionality or co-reference to cases in which one term does not designate anything at all (such as 'phlogiston'). It maps terms of the reduced theory to a variant of the reducing theory that is, given specific conditions, derived from the reducing theory.
(ii) *What are the (primary) candidate-relata for the reduction relation?*
The primary *relata* are theories or, more liberally, scientific representational devices, such as model-descriptions.
(iii) *What is the suggested link to explanation?*
There is no direct link to explanation; however, the idea is that reductions should go together with different kinds of unification, with explanation of the reduced theory, and with the explanation of why the reduced theory worked as well as it did.

The role identity plays in these models is similar to the role it plays in Schaffner's model. First of all, cross-theoretic identities form a limiting case. Moreover, for monistic worlds, this limiting case will be the relevant case. Accordingly, cases of reduction relevant for the monist will be identity-cases of this model. Reduction as based on cross-theoretic identities is required to make sense of at least some reductions that go together with correction. The way correction is conceived in New Wave reductionism is captured by Fig. 7.2:

What is the relation between the replaced theory and its analog, the relation marked with a question-mark in Fig. 7.2? Again, for some cases, it will be a

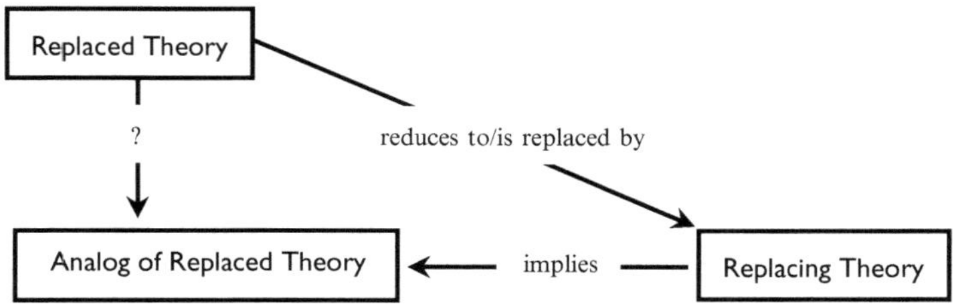

Fig. 7.2 New wave reduction

relation that is astonishingly similar to that of the identity-case of theory reduction. Intuitively, the replacing theory is adjusted to the replaced theory via the analog. The analog of the replaced theory should be derivable from the replacing theory, given specific boundary conditions. This can be thought of as *contextualization* via conditions such as limiting assumptions, so as to capture corrections required to carry out, or to prove the reduction. The relation between the replaced theory and its analog is cashed out in terms of a pairing of expressions, or as a function mapping the terms of the replaced theory to terms of the analog of the replacing theory. *Some* mapping of terms will always be available; so, when is it an interesting mapping function? One way to be an interesting mapping function is to be a mapping function that mimics the fact that the properties purportedly designated by the terms of the replaced theory figure among the properties purportedly designated by the terms of the analog of the replaced theory. The analog relation covers the identity-case. Thus, understanding the analog relation involved in the characterization of the model of new-wave reduction requires understanding the identity case of theory reduction. This will be the case even for some reductions involving substantial correction. If the correction involved in a new wave reduction is achieved by the construction of the analog of the reduced theory from the reducing theory, we will, again, have a case where the mapping of terms is backed up by identity statements (note that these might turn out to be counterfactual: if the reduced theory and its analog were correct, then the relevant identity statements would be true). So, not only in the case of straightforward reductions, but also in some cases involving correction, new wave reduction covers identity-based reduction. To the extent that the analogs introduced *fully account for* the bumpiness of a reduction of a replaced theory to the replacing theory, the analog relation between the reduced theory and its analog will ensure smooth reduction of the replaced theory to the analogue of the replacing theory. Hence, an understanding of how the identity-case of theory reduction works is, again, required for an understanding of how correction works within the new wave model. And again, this model comprises (among other relations) a relation that reconciles diversity and directionality with strong unity, without relying on elimination. For monistic worlds with committal high-level sciences, this form of reduction will be the relevant form, even on the New-Wave conception of reduction. At the same time, this conception covers cases that do rely on elimination.

We have thus sketched the four most prominent suggestions for defining reduction in the philosophy of science. All these characterizations at least cover a case of identity-based reduction. However, it is not obvious how they should fit the explication proposed in Chaps. 4 and 5: all these characterizations are characterizations of a relation the primary *relata* of which are theories, or other scientific representational devices, rather than properties or other non-representational worldly objects. The next section offers a notion of theory reduction that is derivative on the notion explicated above. In Chap. 8 it will be argued that in fact, notions of theory reduction are derivative on notions of property reduction, that is: it will be shown that the procedure to obtain a notion of theory reduction from a notion of ontological reduction proposed in the next section is in fact adequate.

7.6 Theory Reduction

According to the job-description for 'reduction', a fruitful explication should allow for the reduction relation to take properties as its *relata*. Reference to the fact that the consensus in the philosophy of science is, or used to be, that reduction is an inter-theory relation does not all by itself reveal that versions of "ontological" reduction are mistaken. Rather, the burden of proof rests with those who deny that, in the primary sense of the term, reduction is a relation instantiated by pairs of properties. We should be cautious when it comes to introducing restrictions on the possible applications of the reduction-predicate: If we were able to characterize reduction in a way which allows for reductions of properties, we would be able to capture an important aspect of the pre-explicated use of the term 'reduction'. This would be an advantage. This section suggests that based on the explication proposed in Chaps. 4 and 5, we can define a notion of theory reduction and, thus, get all at once, building on the same resources: theory reduction as well as ontological reduction within a coherent framework. The next chapter argues that this conception of reduction is to be preferred over its rivals, which explicate reduction primarily in terms of theory-reduction.

To get an idea of how to connect the explication proposed above to theory reduction, consider again Nagelian reduction: Properly construed, it consists in the idea that the reduced theory should not only be derivable from the reducing theory, but should also cover phenomena that are to be explained in terms of the reducing theory, and connect to the latter by way of bridge-laws that state ontological links.

The idea is this: We take reduction statements concerning links between *kinds* as *basic*, and then characterize how such reduction-statements link the reduced theory to the reducing theory, such that Nagelian reduction is achieved. Thus, these statements about reductive links between the respective kinds (under property structures) play a role similar to bridge-principles in the Nagelian account. Reduction, as explicated above, implies identity. Thus, Nagelian bridge-principles are implied by the corresponding reduction statements. Note that for replacements involving correction, this characterization should be regarded as holding between corrected or refined versions of theories and their counterparts which are formulated in a different vocabulary: This was part of the result of the discussion of Schaffner's model and New-Wave models of reduction.

Here is the definition I suggest (assuming that theory-reductions are effected iff the laws are reduced, and that for laws to be reduced is for them to have their kinds reduced – this relies on the Nagelian picture):

(Explication IX – Theory-Reduction): *T_R reduces to T_B iff for any kind-term t of T_R, there is a kind term t* of T_B, such that the designatum of t, when presented under the concept expressed by t, type-reduces to the designatum of t*, when presented under the concept expressed by t*.*

Hence, in a straightforward sense, the reducing theory explains the phenomena of the reduced theory: It reveals the constitutive structure of entities captured by concepts of the reduced theory.

Unlike rival accounts of theory reduction, which have to take the directionality to stem from features of theories as wholes, we can account for the directionality of reduction right from the beginning: Bridge-principles themselves, thus construed, provide the means to account for the directionality, because the terms connected in such a law relevantly differ in meaning. Replacing the identity sign by the reduction predicate in what functions as bridge-principles thus gives us the resources to account for the directionality of *theory*-reduction, which, on this account, turns out to be a side effect of sets of *ontological* reductions. This has, at least, an intuitive advantage: If our folk-theory about water reduces to chemistry, then it does so partly *because* water, when presented under the folk-concept of water, reduces to water, when presented under the chemical concept of H_2O. Note that this does not contradict the thought that we may *find out* that water reduces to H_2O by discovering the relation between the respective theories; even if the sole justification we could possibly have for believing that water reduces to H_2O is that water-theory reduces to H_2O-theory, this does not imply that it is not the case that water-theory reduces to H_2O-theory *because* water reduces to H_2O. Justification and explanation may run in different directions.

According to this definition, we can define theory reduction in terms of derivation, once we describe the bridge-principles in terms of reduction statements as described above. Let $\{B\}^{TR/TB}$ be the set of bridge-principles of the form of reduction statements as defined above connecting any kind-term of T_R with a kind-term of T_B. Then:

(Explication X – Revised Nagel-Reduction): *T_R reduces to T_B iff*

(i) $T_R \neq T_B$, &
(ii) T_B, $\{B\}^{TR/TB} \vdash T_R$.

We should add that the bridge-principles should be relevantly used in the derivation; thus, they are not redundant. Based on this notion, we can also explicate derivative notions of partial reductions in the sense of reductions that concern fragments of theories only (this is the sense in which Schaffner (2006) employs the term): we just restrict what is said to derive from the base theory and the bridge-laws in (ii) to a subset of the reduced theory.

Recall the characterization of identity-based theory-reduction:

(Principle Identity-Based Theory-Reduction): If T_R reduces to T_B, then the (relevant part of) the ontology of T_R belongs to the ontology of T_B.

If the kind-terms of the respective theories determine a theory's ontology, then we have defined a notion of identity-based theory-reduction. The next chapter argues that this is how we should conceive of theory reduction.

7.7 Conclusion

The explication proposed in the first part of this book can be used to characterize a notion of theory reduction; this is a notion of theory reduction that is based on cross-theoretic identities. It has been shown that the identity-case of theory reduction is covered by a number of prominent models of reduction in the philosophy of science. The explication thus sheds light on the identity-case of theory reduction, which is crucial for an understanding of Nagelian, Structuralist and New-wave models of reduction. We are still here:

Q2: How can this explication be further motivated?

...

Th. 6: It sheds light on the reduction debate in the philosophy of science.

...

The accounts presented thus far all allow for straightforward reduction, or reduction based on cross-level identities, at least as a limiting case. This is why they became disjunctive (Schaffner) or described varieties of reduction in terms of a continuum of relations (New Wave). Moreover, models working with *correction* depend upon a sufficiently precise understanding of how identity-based reductions work: In cases of new wave reduction, in which the job of correction is done by constructing the analog in terms of the reducing theory, we get a sort of reduction that is covered by the proposal offered here. Similarly for Schaffner's model: The corrected reduced theory reduces to the reducing theory just like theories do in the Nagel-picture. Replacement-issues do not necessarily play any role as far as these pairs of corrected and reduced or reducing theories are concerned. Once we correct, say, psychology such that it neatly matches neuroscience, we would get the required identity links, even if uncorrected psychology is false. Similarly, in the case of neuroscience: An adjusted neuroscience may neatly match psychology in its current state. Then, we have reduction of psychology to this variant of neuroscience. Drawing a distinction between these two separate issues, which were not kept separate in the debate, enables us to tell important aspects from less important ones – important as far as an explication of *reduction* is concerned.

In the post-Nagelian philosophy of science, reduction is primarily (or: officially) concerned with reduction as (mere) replacement – interpretations of identity-based reductions have not moved beyond Nagel's insights. This shift in focus perfectly matches the official aim of modeling or reconstructing actual scientific developments. In New-wave reductionism, reduction is primarily treated as mere replacement, and insights gained by investigation of actual examples of theory-successions shaped the relevant model, while at the same time, the notion of reduction is used to formulate metaphysical positions, namely some sort of *reductionism*.

If we define a notion of theory reduction in terms of ontological reduction (as suggested in Sect. 7.6), we capture the identity case *and* can, at the same time, account for directionality. An argument for this suggestion is called for, given what appears to be the consensus in the philosophy of science: that notions of ontological reduction are either confused or can be characterized in more fundamental terms of theory reduction. This point becomes more pressing once we recognize that the puzzle discussed in Chap. 3 seems to vanish, once we define reduction in terms of theory-reduction, referring to holistic criteria of entire theories or models in order to account for the directionality. This is the topic of the next section: It will be argued that in fact, notions of ontological reduction are more fundamental than notions of theory reduction. The account proposed here thus not only sheds light on core notions of reduction in the philosophy of science; it sheds more light on these notions than its rivals.

Bibliography

Balzer, Wolfgang. 1984. On the comparison of classical and special relativistic space-time. In *Reduction in science: Structure, examples, philosophical problems*, ed. Wolfgang Balzer, David A. Pearce, and Heinz-Jürgen Schmidt, 331–357. Dordrecht: Reidel.

Balzer, Wolfgang, and C. Ulises Moulines. 1996. *Structuralist theory of science*. Berlin: de Gruyter.

Bickle, John. 1992. Mental anomaly and the new mind-brain reductionism. *Philosophy of Science* 59: 217–230.

Bickle, John. 1998. *Psychoneural reduction: The new wave*. Cambridge, MA: MIT Press.

Bickle, John. 2003. *Philosophy and neuroscience: A ruthlessly reductive account*. Dordrecht: Kluwer.

Block, Ned, and Robert Stalnaker. 1999. Conceptual analysis, dualism, and the explanatory gap. *Philosophical Review* 108: 1–46.

Carnap, Rudolf. 1934. *The unity of science*. London: Kegan Paul, Trench, Trubner, and Co.

Carnap, Rudolf. 1958. Beobachtungssprache und theoretische sprache. *Dialectica* 12: 236–248.

Churchland, Paul M. 1979. *Scientific realism and the plasticity of mind*. Cambridge: Cambridge University Press.

Churchland, Paul M. 1981. Eliminative materialism and the propositional attitudes. *The Journal of Philosophy* 78: 67–90.

Churchland, Paul M. 1985. Reduction, qualia and the direct introspection of brain states. *Journal of Philosophy* 82: 8–28.

Churchland, Patricia. 1986. *Neurophilosophy*. Cambridge, MA: MIT Press.

Craver, Carl. 2007. *Explaining the brain: Mechanisms and the mosaic unity of neuroscience*. Oxford: Oxford University Press/Clarendon Press.

Darden, Lindley, and Nancy Maull. 1977. Interfield theories. *Philosophy of Science* 44: 43–64.

Dizadji-Bahmani, Foad, Roman Frigg, and Stephan Hartmann. 2010. Who is afraid of Nagelian reduction? *Erkenntnis* 73: 393–412.

Endicott, Robert. 1998. Collapse of the new wave. *The Journal of Philosophy* 95: 53–72.

Endicott, Robert. 2001. Post-structuralist angst – Critical notice: John Bickle, psychoneural reduction: The new wave. *Philosophy of Science* 68: 377–393.

Esfeld, Michael, and Christian Sachse. 2007. Theory reduction by means of functional sub-types. *International Studies in the Philosophy of Sciences* 21: 1–17.

Esfeld, Michael, and Christian Sachse. 2011. *Conservative reductionism*. London: Routledge.

Feyerabend, Paul K. 1962. Explanation, reduction, and empiricism. In *Scientification explanation, space and time, Minnesota studies in the philosophy of science III*, ed. Gover Maxell and Herbert Feigl, 28–97. Minneapolis: University of Minnesota Press.

Feyerabend, Paul K. 1966. 'The structure of science', review of *the structure of science*, by Ernest Nagel. *British Journal for the Philosophy of Science* 17: 237–249.

Feyerabend, Paul K. 1981. *Realism, rationalism and scientific method. Philosophical papers*. Cambridge: Cambridge University Press.

Fodor, Jerry A. 1974. Special sciences: Or the disunity of science as a working hypothesis. *Synthese* 28: 97–115.

Fodor, Jerry A. 1981. *Representations: Philosophical essays on the foundations of cognitive science*. Cambridge, MA: MIT Press.

Friedman, Kenneth. 1982. Is intertheoretic reduction feasible? *The British Journal for the Philosophy of Science* 33: 17–40.

Hempel, Carl G. 1966. *Philosophy of natural science*. Englewood Cliffs: Prentice-Hall.

Hempel, Carl G., and Paul Oppenheim. 1948. Studies in the logic of explanation. *Philosophy of Science* 15: 135–175.

Hoering, Walter. 1984. Anomalies of reduction. In *Reduction in science: Structure, examples, philosophical problems*, ed. Wolfgang Balzer, David A. Pearce, and Heinz-Jürgen Schmidt, 33–50. Dordrecht: Reidel.

Hooker, Clifford. 1981. Towards a general theory of reduction. Part I: Historical and scientific setting. Part II: Identity in reduction. Part III: Cross-categorial reduction. *Dialogue* 20: 38–59, 201–236, 496–529.

Hoyningen-Huene, Paul. 1989. Epistemological reductionism in biology: Intuitions, explications, and objections. In *Reductionism and systems theory in the life sciences: Some problems and perspectives*, ed. F. Wuketits and P. Hoyningen-Huene, 29–44. Dordrecht: Kluwer.

Hull, David L. 1974. Informal aspects of theory reduction. In *PSA: Proceedings of the biennial meeting of the philosophy of science association*, ed. G. Pearce, A.C. Michalos, C.A. Hooker, and R.S. Cohen, 653–670. Dordrecht: Reidel.

Kemeny, John G., and Paul Oppenheim. 1956. On reduction. *Philosophical Studies* 7: 6–19.

Kim, Jaegwon. 1993. *Supervenience and mind. Selected philosophical essays*. Cambridge: Cambridge University Press.

Kim, Jaegwon. 1998. *Mind in a physical world. An essay on the mind-body problem and mental causation*. Cambridge, MA: Harvard University Press.

Kuhn, Thomas. 1962. *The structure of scientific revolutions*. Chicago: University of Chicago Press.

Marras, Ausonio. 2005. Consciousness and reduction. *The British Journal for the Philosophy of Science* 56: 335–361.

Moulines, C. Ulises. 1984. Ontological reduction in the natural sciences. In *Reduction in science: Structure, examples, philosophical problems*, ed. Wolfgang Balzer, David A. Pearce, and Heinz-Jürgen Schmidt, 51–70. Dordrecht: Reidel.

Nagel, Ernest. 1935. The logic of reduction in the sciences. *Erkenntnis* 5: 46–52.

Nagel, Ernest. 1949. The meaning of reduction in the natural sciences. In *Science and civilization*, ed. Robert Stouffer, 99–135. Madison: University of Wisconsin Press.

Nagel, Ernest. 1961. *The structure of science. Problems in the logic of explanation*. New York: Harcourt, Brace & World, Inc.
Nagel, Ernest. 1970. Issues in the logic of reductive explanations. In *Mind, science, and history*, ed. Howard Kiefer and Milton Munitz, 117–137. Albany: SUNY Press.
Nickles, Thomas. 1973. Two concepts of intertheoretic reduction. *The Journal of Philosophy* 70: 181–201.
Oppenheim, Paul, and Hilary Putnam. 1958. The unity of science as a working hypothesis. In *Concepts, theories, and the mind-body problem*, ed. Grover Maxwell, Herbert Feigl, and Michael Scriven, 3–36. Minneapolis: Minnesota University Press.
Putnam, Hilary. 1965. How not to talk about meaning. In *Boston studies in the philosophy of science, vol. 2, in honor of Philip Frank: Proceedings of the Boston Colloquium for the philosophy of science*, 1962–1964, ed. Robert S. Cohen and Marx W. Wartofsky, 205–222. New York: Humanities.
Sachse, Christian. 2007. *Reductionism in the philosophy of science*. Frankfurt/M: Ontos-Verlag.
Salmon, Wesley. 1989. Four decades of explanation. In *Scientific explanation: Minnesota studies in the philosophy of science, 13*, ed. Philip Kitcher and Wesley Salmon, 3–219. Minneapolis: University of Minnesota Press.
Sarkar, Sahotra. 1992. Models of reduction and categories of reductionism. *Synthese* 91: 167–194.
Schaffner, Kenneth. 1967. Approaches to reduction. *Philosophy of Science* 34: 137–147.
Schaffner, Kenneth. 1969. Correspondence rules. *Philosophy of Science* 36: 280–290.
Schaffner, Kenneth. 1974. The peripherality of reductionism in the development of molecular biology. *Journal of the History of Biology* 7: 111–139.
Schaffner, Kenneth. 1976. Reductionism in biology: Prospects and problems. In *Proceedings of the 1974 meeting of the Philosophy of Science Association*, 1974, ed. A. Michalos and R.S. Cohen, 613–632. Dordrecht: Reidel.
Schaffner, Kenneth. 1993. *Discovery and explanation in biology and medicine*. Chicago: The University of Chicago Press.
Schaffner, Kenneth. 2006. Reduction: The Cheshire Cat problem and a return to roots. *Synthese* 151: 377–402.
Schaffner, Kenneth. 2012. Ernes Nagel and reduction. *Journal of Philosophy* 109: 534–565.
Scheibe, Erhard. 1999. *Die Reduktion Physikalischer Theorien: Inkommensurabilität und Grenzfallreduktion*, vol. II. Berlin/Heidelberg/New York: Springer.
Silberstein, Michael. 2002. Reduction, emergence, and explanation. In *The Blackwell guide to the philosophy of science*, ed. Michael Silberstein and Peter Machamer, 80–107. Oxford: Blackwell.
Sklar, Lawrence. 1967. Types of inter-theoretic reduction. *British Journal for Philosophy of Science* 18: 109–124.
Sneed, Joseph. 1971. *The logical structure of mathematical physics*. Dordrecht: Reidel.
Stegmüller, Wolfgang. 1979. *The structuralist view of theories. A possible analogue of the Bourbaki programme in physical science*. Berlin/Heidelberg/New York: Springer.
Stegmüller, Wolfgang. 1986. *Theorie und Erfahrung: Dritter Teilband. Die Entwicklung des neueren Strukturalismus seit 1973*, vol. 3. Berlin/Heidelberg/New York: Springer.
Suppes, Patrick. 1967. What is a scientific theory. In *Philosophy of science today*, ed. Sidney Morgenbesser, 55–67. New York: Basic Books.
van Riel, Raphael. 2011. Nagelian reduction beyond the Nagel-model. *Philosophy of Science* 78: 353–375.
Wimsatt, William C. 1974. Complexity and organization. In *PSA: Proceedings of the biennial meeting of the philosophy of science association*, 1972, *Boston studies in the philosophy of science*, ed. Kenneth Schaffner and R. Cohen, 67–86. Dordrecht: Reidel.

Chapter 8
Theory Reduction and Holism

Several distinct explications of reduction may be appropriate for distinct uses of the term. The present chapter argues that in several important respects, the explication laid out in the first part of this book is more fundamental than any explication of reduction that explicates the notion in terms of an inter-theory relation. Prima facie, though, if the primary *relata* of the reduction relation were theories, we would immediately get rid of a number of problems all at once. Amongst other things, we could account for directionality by reference to differences between reduced and reducing theories, concerning features such as the degree of unification, or the respective theories' explanatory power. The present section argues that an explication of reduction along these lines is derivative on a notion of ontological reduction.

8.1 A Sketch of the Alternative

If the primary *relata* of the reduction relation are theories, we get rid of a number of problems all at once. First, the puzzle discussed in Chap. 3 vanishes: Unity stems from identity of kinds in theory reductions. Diversity is easily accounted for in terms of differences between reduced and reducing theories. Hence, no puzzle lurks here. Moreover, we may hope to describe the directionality in terms of holistic features of theories. Tying reduction to theories seems to be metaphysically innocent, or at least more innocent than tying reduction to kind-identity. So, if there is a cheap solution available, why stick to the costly one, especially if the costly one could be characterized in terms of the cheap one, as, for example, suggested by Nagel and Churchland (Nagel 1961, Chapter 11, Sec. III.3, 1970, 119; Churchland 1986, 278)?

To assess this issue, it will be useful to give the view that theory-reduction is prior to ontological reduction a label. I suggest referring to it as 'the holistic view on reduction', or 'holism' for short. It consists in the idea that either, ontological

R. van Riel, *The Concept of Reduction*, Philosophical Studies Series 121,
DOI 10.1007/978-3-319-04162-9_8,

reduction is just confused or that the truth of statements of reduction relations whose *relata* are kinds or properties some theory employs in its ontology depend upon the truth of statements about reductions of theories or sufficiently complex models. The idea is *holistic*: If water reduces to H_2O, then it does so because, or in the sense that, our folk-theory of water reduces to chemistry as a whole. According to holism, for water to reduce to H_2O, water, or 'water', must be relevantly related to our folk-theory of water, H_2O, or 'H_2O', must be relevantly related to chemistry, and our folk-theory of water must reduce to chemistry. There is nothing more to the reduction of water to H_2O than that. Stating these connections and that our folk-theory about water reduces to chemistry, we explain why and in which (derivative) sense water reduces to H_2O. It is fair to say that in traditional philosophy of science the idea that theory reduction is more basic than ontological reduction, if not the only coherent form of reduction, is common ground.

In contrast, on the view defended here, it is semantic features of expressions that give rise to diversity. Directionality is accounted for in these terms. According to holism, it is standing in a relation to scientific theories that give rise to diversity – any ontological conception of reduction turns out to be merely derivative upon a concept of theory reduction. Similarly, directionality will be accounted for in terms of differences between theories. This view is widespread. But why to come up with the idea that theory-reduction is more basic in the first place?

Many philosophers of science are not particularly enthusiastic about conceptions of *ontological reduction*. A detailed discussion of the motives for this view would take us too far from our actual target, partly because these motives are largely independent from the specific views these philosophers take on reduction. Rather, they depend upon broader conceptions of realism, and the relevance of epistemological and pragmatic assumptions in metaphysical theories.[1] Instead of presenting a detailed argument against this view, I will show that holism fails when it comes to making explicit the commitments that go together with reduction-talk. Sections 8.2, 8.3, 8.4 and 8.5 discuss different holistic candidates to account for directionality. Section 8.6 argues that the semantic and metaphysical commitments that go together with holistic approaches to reduction, according to which ontological reduction derives from theory reduction, are equivalent to the commitments of the "ontological" explication proposed above. In Sect. 8.7 it is argued that this explication is more fundamental than holistic interpretations of theory reduction – an argument parts of which will be further developed in the final Chap. 9.

[1]This seems to be Nagel's point: We do not get access to properties independent of our theories. This is supposed to show that a concept of a relation holding between properties independent of our theories is confused (Nagel 1961, Chapter 11, Sec. III.3, 1970, 119), see, for a discussion, (van Riel 2011).

8.2 Directionality and the Reductive Link

Models of theory-reduction offer a number of different criteria we might want to refer to in order to account for the *directionality* of reduction. A first candidate might be this: *T_R reduces to T_B only if T_B corrects T_R and T_R does not correct T_B.* This would be an option, as long as correction is spelled out in a way that does not (i) cover and (ii) rely on the identity-case, where no correction is involved. Building on the insights gained in Chap. 7, we are safe to conclude that it is, thus, not a *serious* option. Similarly, there is another group of relations referred to in the reduction-debate that do not help, a group we are already familiar with, such as identity and derivability. We should briefly list these candidates. Obviously, they do not make for directionality:

Identity of Ontology	This stems from some pre-theoretic descriptions, the bridge law interpretation in Nagel, identity-theories in the philosophy of mind, it is Moulines' candidate for an amendment of semantic models and it occurs in New-Wave reductionism as a candidate for grounding 'smooth reductions'.
Derivability	This is inspired by Nagel's model.
Logical equivalence	This is the formal characterization of bridge-laws in Nagel's model.
Implication	This is the minimal condition for derivability which is considered by Richardson and Nagel as figuring in bridge-principles.
Structural isomorphism	This and other analog relations underlie new-wave reductionism and structuralist versions of reduction.
Nomological equivalence	This covers weaker syntactic forms of bridge-law interpretations (*cf.* van Riel 2011).

Let us make it short: None of these relations is "directional". When these relations are assigned to '*R*' in the following schema: '*aRb & bRa*', an instance of this schema possibly expresses a truth.[2] Thus, these relations alone will not enable us to account for the directionality. This can be illustrated reflecting on the official Nagel-model. According to Nagel's official definition, (i) every theory reduces to itself, (ii) every theory reduces to every theory which is necessarily false (*ex falsum quodlibet*), and (iii) every pair of theories the elements (sentences) of which imply each other given the same set of bridge-laws reduce to each other. So, as it stands, it cannot be used to account for directionality. Clearly, reference to identity as the

[2]Theo Kuipers nicely argued that "[a] moment's reflection shows that we can reverse the standard argument: from the ideal gas law and KTR [kinetic temperature relation] we can derive the kinetic hypothesis. Hence the reduction seems to work in both directions. Intuitively [...] we are inclined to expect, almost by definition, asymmetry [...]" (Kuipers 1982, 107f). And this is a problem, indeed.

basis for bridge-laws will not help. Here is an additional candidate that is worth mentioning, partly because it points to an interesting feature of the connection between reduction and explanation: Schaffner, referring to Berent Enç (Schaffner 1993, 469; Enç 1976), has suggested that reduction is a *causal* relation.

This idea seems misguided, as also Schaffner tends to suggest (Schaffner 1993, 467f.), although this criterion later enters his tentative list of additional criteria making for the directionality ("asymmetry", in Schaffner's terms) of reduction. I already suggested that the fact that Enç employs *causal* talk to describe what he also calls a 'generative relation' is highly misleading (see Sect. 5.7.2). Discussing the reduction of classical thermodynamics to the kinetic theory of gases, Berent Enç correctly noted that we "cannot explain why the rate of change of momentum of the molecule is such and such by the fact that the gas exerts such and such a pressure on them all" (1976, 296). Since we can explain a given amount of pressure referring to the rate of change of momentum of molecules, he concludes that there must be a (synchronous) *causal* relation between the amount of pressure of gas and the average rate of change per unit area and time of the linear momentum of the gas molecules, making for the relevant asymmetry. This extremely liberal version of causation is supposed to base reductions between theories (1976, 295). Enç consequently interpreted bridge-principles as stating these generative relations (ibid. 296), and he seems to be well aware of the fact that this liberal version of causation should be compatible with co-occurrence, and identity. Although his explications remain rather vague in this respect, he seems to suggest that the relevant difference between the generating fact and the fact so generated is *epistemic* in nature.

For the present purpose, let us, however, discuss his suggestion in terms of causation, as ordinarily construed. This is, at least, the way it is interpreted by Schaffner. This interpretation does not make for a notion of reduction that can be used to illuminate the claim of strong unity, because it is too weak a notion to guarantee monism; a view building on causation in the classical sense is committed to denying the relevant identity-statements. Secondly, it seems plainly false. Let us assume that our ordinary discourse about stuff in the world reduces to chemistry, and that the bridge-laws required for this reduction take a form similar to the prominent example of water being connected to H_2O. Now, H_2O does not *cause* (something to be) water, it just *is* water (if reductionism regarding water is true). If the bridge principles used in the case of thermodynamics and the kinetic theory of gases behave in a similar fashion, then they do not state causal relations. Enç is right in that there is some generative or explanatory relation holding between facts at or propositions about the fundamental level and facts at or propositions about some higher level. We can *explain* pressure by reference to a rate of change of momentum of molecules. But pressure *is* rate of change of momentum of molecules (if the reductionist interpretation of this case is correct), it is not *caused* by it. It makes sense to say that being H_2O makes water *watery*, or that we can explain certain features of water in terms of features of H_2O. Its being H_2O is responsible for its, say, being almost tasteless for ordinary humans. Similarly, an object's having a certain mean kinetic energy is responsible for its having a certain temperature. An object's temperature, however, is by no means the *causal result* of the mean kinetic energy. Temperature and mean kinetic energy occur at the same time in the same objects, and (if the relevant sort of reductionism holds) necessarily so, and temperature and mean kinetic energy have the *same* causal role. So, the explanatory dependence observed by Enç has to be spelled out in different terms.

Thus, none of the relations offered in the literature (and I am not aware of another candidate) adequately captures the "reductive link". None of them makes sense of the directionality of reduction. Instead, we should, in the spirit of Holism, take a look at other features of entire theories that could give rise to the directionality of reduction. Reduction reconciles diversity and directionality with unity. Why not tie diversity to differences between pairs of reducing and reduced theory?

8.3 Directionality and Differences Between Theories

The more elaborate holistic answer to the question of where the directionality of reduction stems from can roughly be put as follows: Reducing theories may vastly differ from their reduced counterparts. Thus, we may hope to find a criterion for the directionality among these differences. To avoid long and complex exegetical remarks, I will just list the possible candidates, mentioning some prominent occurrences of these ideas in the literature. The following two candidates can be obtained from Nagel (1961, *cf.*; van Riel 2011).

(1) The reducing theory should be *better* established than the reduced one in virtue of a wider range of application.
(2) Reduction is concerned with unification: The reducing theory unifies different laws, which seemed unconnected before the actual reduction took place.[3]

These two suggestions can also be found in Churchland (1986). Schaffner briefly addresses the issue of directionality. He suggests that the directionality can be spelled out by reference to the fact that reducing theories deal with (proper) parts of the objects of the reduced theory, and by reference to the idea that the reducing theory will explain much where the old theory was "buffaloed": It has a *greater range of explanation* and, in addition it supposedly explains the domain of the reduced theory, the reduced theory itself, and why the reduced theory worked as well as it did. Add the criteria taken from Churchland (1986, Chap. 7.5), namely, that reduction goes together with explanatory unification and ontological unification, we get seven additional criteria:

(3) Reduction goes together with explanatory unification. (Churchland 1986, 279)
(4) Reduction goes together with ontological unification. (Churchland 1986, 280)

[3]Note that this criterion is epistemic in that it builds upon a difference in what seemed to be the case before the reduction took place, and what does not seem to be the case any more after the reduction took place.

(5) In virtue of a reduction, the range of explanation is expanded. (Churchland 1986, 283; Schaffner 1993, 469; Hooker 1981, 211)
(6) The old theory is explained by the reducing theory. (Churchland 1986, 283)
(7) The new theory explains why the old one worked as well as it did. (Churchland 1986, 283; Schaffner 1993, 429; Sklar 1967; Hooker 2006, 158)
(8) Reducing theories are (usually) concerned with parts of the wholes the reduced theory is concerned with. (Schaffner 1993, 469)
(9) The reducing theory explains the domain of the reduced theory. (Schaffner 1993, 469)

First of all, it is worth noting that these criteria were obtained by empirical investigations of cases of alleged reductions. Protagonists in the debate arrived at them by relying on an *extension first!* approach. Thus, one should not expect these criteria to be "built-into" a predefined notion of reduction we already grasp when engaging in philosophical discourse. Hence, we would be extremely lucky if the cases of reductions we investigate really reveal the relevant conceptual component we need in order to appropriately explicate the concept of reduction. This also explains why the following question, that almost immediately suggests itself, seems appropriate: *Why* do alleged cases of reduction possess these features? What has *reduction* got to do with it? Note that this point has escaped Churchland's attention when, having specified the relevant variant of reduction, she writes:

> Under these conditions the old theory reduces to the new. When reduction is successfully achieved, the new theory will explain the old theory, it will explain why the old theory worked as well as it did, and it will explain much where the old theory was buffaloed (1986, 283)

Churchland seems to suggest that the directionality *derives from* the concept of reduction: I take this to suggest that *in virtue* of the reductive link, we will expect the reducing theory to explain the old one and so forth. Thus, these explanation-relations enter the game *in virtue of reduction*, such that we should not expect the reduction model to be (illuminatingly) definable via *effects* of the instantiation of a reduction relation. Based on the non-holistic model of reduction developed in Chap. 5, we will be in a position to argue against holistic approaches as follows: On the non-holistic account developed here, we can in fact *account for why we should expect at least some cases of reductions to exhibit these features*. On the holistic account, these features are left unexplained. An explanation is called for, and any notion of reduction that yields such an explanation is more fundamental than the holistic one.

Before turning to this argument in Sect. 8.7, let me, firstly, argue that conditions (6) and (9) miss the target (8.4). In a second step, I will show that *all* of these criteria fail, if possibly, there exist relevantly equivalent theories that may instantiate the reduction relation (8.5). Thirdly, I will show that even if we were able to come up with a coherent holistic account of reduction, we would still have to account for reduction in terms of exactly those semantic and metaphysical properties we have referred to within the "ontological" account proposed in the first part of this book. Thus, holistic approaches do not come cheaper in this respect (8.6).

8.4 Two Problems for Holistic Approaches to Theory-Reduction

According to (6), the reducing theory explains the reduced theory – a rather dubious formulation that needs to be interpreted. Under one interpretation, this condition seems to be tied to the idea of reduction as replacement. In this sense, it is an explanation that at the same time shows why the reduced theory worked as well as it did in some respects even though it was, in some other respects, 'buffaloed'. For example, oxygen theory gives the resources to explain why phlogiston-theory was able to account for why, say, candles dye out when covered by a cub; it did, because in this respect, both theories *yield similar results*. Referring to oxygen theory, we can explain why the regularities of phlogiston theory hold. In this sense, condition (6) can be ignored, since we are not concerned with *correction* or *mere replacement*. But isn't there a sense in which a theory explains another theory "directly"? Not in any straightforward sense that does not immediately collapse into the model of reduction proposed above. Conceive of a theory as a conjunction of sentences. Let T_R stand for the reduced theory and T_B for the reducing theory. This is what *direct* theory explanation seems to consist in (schematically): 'T_R because T_B'.[4] However, in this case, the truth of this explanation will depend upon explanations in which individual conjuncts of the reduced theory are explained by individual conjuncts of the reducing theory, just like the model of theory reduction that depends on ontological reduction proposed here would have it. Consider the following example: water freezes below 0 °C under normal conditions (p_1) because H_2O molecules form lattice structures below 0 °C under normal conditions (q_1). Now, take some other explanation: p_2 *because* q_2 (so that the conjunction of p_1 and p_2 and the conjunction of q_1 and q_2 can be conceived of as dummies for theories). Combining these explanations, we arrive at (p_1 & p_2) *because* (q_1 & q_2), which seems to depend on the truth of the individual explanations we started with. This seems to be the sort of connection we would get in "direct" theory explanation. At least in this case, the individual explanations, guaranteed by the model proposed here, are more fundamental.

Alternatively, one might suggest that a reducing theory *can be used* to explain why the reduced theory worked as well as it did (this is Schaffner's interpretation of theory explanation (Schaffner 1993, 429)). Thus, under an appropriate interpretation, (6) just states what (7) states. Assuming that no other interpretation is available, (6) either collapses into (7), into the model proposed here, or is misplaced in an explication of reduction.

---

[4]Recall *(Explication III – Facts)*: The fact that p reduces to the fact that q *iff* (i) p because q, and (ii) the fact that $p =$ the fact that q. An anonymous referee pointed out that we might conceive of one version of theory reduction in terms of the reduction of the fact that the reduced theory is true to the fact that the reducing theory is true. Depending on how fine-grained we individuate facts and truth-bearers, this might fit the idea of fact reduction, so that theory-reduction just is, or can be conceived of as a specific form of fact reduction. The reduced theory is true *because* the reducing theory is true, and the fact that the reduced theory is true just is the fact that the reducing theory is true.

What about criterion (9), according to which the reduced theory's domain is explained by the reducing theory? As it stands, it does not help; if the reduced theory explains its domain, too, then this criterion does not make for directionality. The next section considers the idea that there is a possible case of reductions that is not covered by any of the candidates (1)–(9).

8.5 Holism and the Problem of Equivalent Theories

To give a conceptually (and, maybe, metaphysically) possible example of a reducing and a reduced theory, which are equivalent with respect to conditions (1)–(9), assume that these theories describe a world which is inhabited by super-sensitive cognitive systems. These systems are super-sensitive in the sense that for any neural difference, there is an experiential, phenomenal difference. In addition, in this world, psychology reduces to neuroscience. The generalizations folk-psychology comes up with are equivalent to the generalizations neuroscience suggests: neuroscience, in this possible world, is a complete vindication of psychology. For this to be possible, these theories should meet the requirements needed for cross-level links that are based on identity of the referents of terms that designate kinds. Depending on which stance we take towards disjunctive kinds or cross-level identification of contextualized kinds (see above, Sect. 6.5), we can tell different stories about how these requirements could be met. If we need contextualized kinds to yield cross-level identification, these beings' brains should lack plasticity, the theories should contain species-specific expressions for mental properties and so forth.

Assume that our super-sensitive beings exemplify three basic neural properties N1, N2, and N3. Assume also that there are three basic phenomenal properties, P1, P2, and P3. Assume that N1 occurs iff P1 and P2 occur, N2 occurs iff P2 and P3 occur, and N3 occurs iff P1 and P3 occur. Assume that N1 and N2 occur iff P1 occurs, N2 and N3 occur iff P2 occurs, and N1 and N3 occur iff P3 occurs. Moreover, assume that the following identity statements are true:

The type instances of which instantiate N1 & N2 = P1
The type instances of which instantiate N2 & N3 = P2
The type instances of which instantiate N1 & N3 = P3
The type instances of which instantiate P1 & P2 = N1
The type instances of which instantiate P2 & P3 = N2
The type instances of which instantiate P1 & P3 = N3

For this to be possible, the phenomenal states P1 and P2 can occur even though the condition for P1 to occur in isolation is not fulfilled. Thus, N1 gives rise to P1 and P2 only if N1 occurs in isolation (otherwise, we would run into the problem that given one state, these beings would be in every state they are possibly in). Assume that there is just one law in psychology and one law in neuroscience. Let [B] present some behavioral output. For neuroscience, we get:

A state that instantiates N1 and N2 occurs → B occurs

Consequently, the psychological the law will be:

A state that instantiates P1 occurs → B occurs

In this case, moving from psychology to neuroscience will not increase explanatory power. Any difference we can account for in neuroscience is a difference we can account for in psychology. However, it might nevertheless be the case that there is some reductive move from, say, the type instances of which instantiate P2 & P3 to the type N2. Why is that? Assume that the phenomenal character of the concepts employed in psychology is responsible for the fact that the relevant reductive links are instantiated. Or assume that the psychological concepts are functional concepts, where psychological states are tied to some input and behavioral output, whereas the concepts employed in neuroscience are physiological in nature. Or assume that there are no psychological concepts, but rather directly referring terms. Again, under these circumstances, psychology seems to reduce to neuroscience.[5]

Now, let us turn to the holistic criteria, none of which is applicable in the present case:

(1) The reducing theory should be *better* established than the reduced one in virtue of a wider range of application.
(2) Reduction is concerned with unification: The underlying theory unifies different laws which seemed unconnected before the actual reduction took place.
(3) Reduction goes together with explanatory unification.
(4) Reduction goes together with ontological unification.
(5) In virtue of reduction, the range of explanation is expanded.
(6) The old theory is explained by the reducing theory.
(7) The new theory explains why the old one worked as well as it did.
(8) Reducing theories are (usually) concerned with parts of the wholes the reduced theory is concerned with.
(9) The reducing theory explains the domain of the reduced theory.

---

[5]The modality involved in this argument has a conceptual and a metaphysical reading. According to the metaphysical reading, it can be shown that any explication of the concept of reduction fails to cover some metaphysically possible cases of reduction. According to the conceptual reading, the explication misses some of the conceptual features of reduction, even though the *relation* picked out by the definition might be what philosophers had in mind. However, a conceptually appropriate definition which blocks these conceptually possible examples will be advantageous over a definition which does not block these conceptually possible examples. Given some ontological dispute – like, say, the possibility of zombies – mere conceptual possibility is maybe not enough to prove the intended metaphysical possibility. Here, however, we are concerned with issues of definition, and, thus, with conceptual issues.

By assumption, both theories are equally well established, the reducing theory does not unify laws, it does not unify by explanation, nor does it unify some other theories' ontologies; the range of explanation is not expanded, the old theory is not explained by the reducing theory (or if it is, then it is in a derivative sense), and the new theory does not explain why the old theory worked as well as it did. No part-whole relation is involved. (9) is true, but it does not help, because the reducing theory explains the domain of the reduced theory just like the reduced theory explains the domain of the reducing theory. Thus, if the thought-experiment just presented is telling, the criteria which are supposed to suit holism miss the target. On the explication presented above, the reduction in the present case is straightforward, if the concepts employed by this fictional neurophysiology are more transparent to the kinds' natures than the psychological concepts.

However, one might still tend to think that reference to semantic facts about expressions or their conceptual features is something we are better off without. As long as holistic models do not have to appeal to things as strange as Fregean modes of presentations, or conceptual contents to make sense of reduction, we should bite the bullet: We should accept that we have to slightly change the concept of reduction (such that possible cases as the one just presented are not covered – but so what? They are hardly relevant for the actual progress of science!), but still have the advantage of getting rid of these strange semantic notions. Let us thus move on to an argument for the claim that holistic approaches have to build upon similar features as well and are, thus equally expensive.

8.6 Holism and the Problem of "Ontological" Reductions

The idea of holism that accounts for "ontological" reduction in terms of theory reduction is this: If '*a* reduces to *b*' expresses a truth, then it does so because '*a*' is relevantly related to a theory *t* and '*b*' is relevantly related to a theory *t**, and *t* reduces to *t**. I will now argue that for a term '*a*' to be relevantly related to a theory is for it to have specific semantic properties other than reference (namely, those that are relevant for reduction as construed above), that are relevantly related to *t*. Thus, it follows that those who embrace this form of holism are committed to the assumption that the relevant semantic facts play a crucial role in reduction-statements. Since the accounts of theory reduction we are familiar with are, on the identity reading, also committed to the truth of the corresponding metaphysical claims (those expressed by or rendering true bridge-laws, or statements about analog relations or isomorphisms), holism based on these accounts does not come cheaper than the account presented above.

Here is a test: We evaluate a complex holistic reduction-statement and check whether or not we can change *meaning* (the relevant semantic facts) while preserving *spelling* and *reference salva veritate*. Thus, we keep the relevant expression stable, and we keep the referent of the expression that is supposed to be relevantly related to the reducing or reduced theory stable, and vary meaning of this expression, thereby showing that varying these semantic facts, we disconnect the expression from the reduced or reducing theory. If so, the truth of holistic reduction statements depends on the truth of statements about connections between conceptual contents and theories. 'Water' and 'H_2O' are, on the holistic approach, *relevantly related to* our folk theory of water and chemistry in virtue of their conceptual content.

To see this, consider the expression 'ferrum'. Insofar as the term expresses the concept it expressed in ordinary language *Latin*, it is relevantly related to what was folk chemistry back then. Insofar as the term expresses the concept it expresses in current chemistry, also captured by 'Fe', it is relevantly related to current chemistry. Reference or designation is, or so we shall assume, the same in both uses, just like spelling.

Holism takes for granted that a reduces to b only if $a = b$, and 'a reduces to b' to be derivatively true (true in virtue of the corresponding theory-reduction). The 'ferrum' case allows for the following construction: It is one of the rare cases in which 'a reduces to b' and "'a' = 'b'" both express a truth, that is, the instances of 'a' and 'b' are identical. Let us assume that *ferrum*, (as used in folk-chemistry in Latin), reduces to *ferrum* (as used in chemistry). Then, *ferrum* (as used in folk-chemistry in Latin) = *ferrum* (as used in chemistry), and, obviously, 'ferrum' = 'ferrum'. So, why is, then, 'ferrum', as used in Latin, relevantly related to folk-chemistry, rather than chemistry? This is partly due to semantic facts other than reference, such as being a directly referring expression, or being associated with the relevant phenomenal properties, or being illuminatingly definable by the appropriately modified Ramsey-Sentence of folk-chemistry. 'Ferrum', as used in chemistry, does not exhibit these features. This is why it is relevantly related to chemistry, rather than folk-chemistry.

These cases are rare because, fortunately, semantic differences between ordinary language expressions and their allegedly co-referential counterparts in science are usually made transparent by lexical differences. The truth of sentences of the form "'a' is relevantly related to theory t', when used to capture the relevant relation between an expression, such as 'water', and the corresponding theory, such as our folk-theory of water, depend upon facts about the meaning of what takes the position of 'a'. Therefore, any holistic approach to reduction, which has to rely on an expression's being relevantly related to a theory, has to account for the relevant semantic features of the candidate-expressions.

Thus, holism is as expensive as the non-holistic account offered here.[6] Taking into account that at least in the philosophy of mind, there is an established use of

[6]Now, it should be obvious that if the meanings of theoretical terms of a given theory are fixed by the role they play within this theory, then, in a sense, for reductions that work with theoretical terms, *some* sort of holistic interpretation is required; the relevant theory defines the meaning of the theoretical term (and from that theory, we obtain the term's meaning) and in virtue of playing the role it does within that theory, it may occur in the relevant reduction statements (because it occurs in some reduction-statements partly in virtue of the meaning it has). This is, however, not to be conflated with the idea that for any reduction statement connecting kind terms, *theory* reduction

'reduction' in the "ontological" sense, and having shown that there is a perfectly intelligible way of explicating this idea, I hope to have persuaded the reader that the path pursued here is not obviously a dead end. The next section argues that in fact, the explication proposed above is more fundamental than theory reduction, in that the candidate criteria to account for the directionality of reduction in holistic terms can be *explained* referring to this explication. These criteria primarily stem from observations of alleged cases of reductions. Insofar as we can account for these observations within the framework proposed here, this framework turns out to illuminate what is common to a number of reductions and, thus, turns out to be fruitful in this respect.

8.7 The Priority of the Explication and the Failure of Holism

In Chap. 5, it has been shown that the explication proposed here perfectly matches the concept's job-description. Amongst other things, it enables us to straightforwardly account for paradigmatic cases of reduction, such as the reduction of pain to C-fiber activation, and it offers the resources to characterize various derivative notions of reduction such as partial reduction, plural reduction, type-reduction, token reduction and theory-reduction. Still, we lack an argument to decide what is more fundamental: A notion of theory reduction or the "ontological" notion explicated above. This section suggests that the explication proposed above is more basic than explications in terms of theory-reduction. It is more basic in the sense that features of models of theory reduction, which allegedly give rise to directionality, depend on the semantic and explanatory facts alluded to in the explication proposed here. Recall these suggested additional features referred to in definitions of theory-reduction:

(1) The reducing theory should be *better* established than the reduced one in virtue of a wider range of application.
(2) Reduction is concerned with unification: The underlying theory unifies different laws which seemed unconnected before the actual reduction took place.
(3) Reduction goes together with explanatory unification.
(4) Reduction goes together with ontological unification.
(5) In virtue of reduction, the range of explanation is expanded.
(6) The old theory is explained by the reducing theory.
(7) The new theory explains why the old one worked as well as it did.

---

is required. First of all, these statements are just different. Secondly, to understand the first, you do not have to understand the second. Thus, it is implausible to assume that the former be conceptually dependent upon the latter (and vice versa: you can believe that a theory reduces to another theory without even having the notion of meaning holism). Thus, one can accept meaning-holism for theoretical terms without thereby being committed to the assumption that individual reduction statements are dependent upon theory-reduction statements.

(8) Reducing theories are (usually) concerned with parts of the wholes the reduced theory is concerned with.
(9) The reducing theory explains the domain of the reduced theory.

In the literature, these are offered as additional features of an appropriate characterization of reduction. This is to a certain extent dissatisfying: As already mentioned, instead of *stipulating* that reductions should conform to these criteria, which are usually inspired by careful examinations of alleged cases of reductions, one might want to know *why* cases of reduction exhibit these features, if they do. An explication of reduction that enables us to answer this question will meet a crucial adequacy condition: to the extent that it illuminates these features, the explication is fruitful.

It has been argued that there are possible cases of reductions for which all these criteria fail (see Sect. 8.5). Moreover, conditions (6) and (9) have been rejected. They do not form interesting candidates for describing directionality (8.4). It should be obvious that on the explication offered here, we can explain why reduction amounts to the fact that the reduced theory's phenomena are explained by the reducing theory. In one sense, this is the consequence of identification; the ontology of the reduced theory is subsumed under the ontology of the reducing theory. Some of the other criteria can be accounted for in a similarly straightforward way. The next sections deal with conditions (1) and (5)–(9). A discussion of conditions (2)–(4), which heavily rely on the notion of unification, will have to wait until Chap. 9.

8.7.1 *Reduction and Explanation in Terms of Parts*

Consider (8), according to which reduction is tied to part-whole relations: Insofar as the constitutive structure of entities is to be cashed out in terms of parts of these entities and their properties, as it is in the case of water and H_2O, reduction is concerned with part-whole relations. This, however, is not necessarily the case, as examples of 'Iron' and 'Fe', or 'pain' and 'C-fiber-activation' reveal (see Sect. 5.5). Insofar as our world is such that wholes behave the way they do or are what they are in virtue of their parts, reduction is concerned with parts. However, (i) there are cases where this parallelism brakes down, and (ii) we should not build this *empirically interesting result* into the notion of reduction. Recall: The idealist might use the term 'reduction' just like the materialist, without contradicting herself. Thus, building on the explication of reduction offered here, we can explain why, if materialism is true, reduction *often, though not essentially,* amounts to a re-description of entities in terms of their parts, and can thus explain why, given materialism, we should expect some actual cases of reduction to accord with condition (8).

8.7.2 Reduction and a Theory's Range of Application

Consider (5) (and, in the course of this discussion, (9) and (1)): Insofar as we re-describe an entity in terms of its constitutive structure, we get access to why the entity behaves the way it does. The constitutive structure is responsible for the object's behavior. Salt dissolves in water. But why? This is so in virtue of the chemical properties of salt and water. These are made accessible when the relevant entities are adequately re-described in terms of constitutive property structures. An object's nature is not only what determines what it is, but also what determines its behavior. Providing a constitutive structure of an entity, we get access to properties that are responsible for its surface properties. A *theory* that describes an object in terms of its constitutive structure contains laws or law-like statements about connections between the properties that constitute the property-structure under which the entity is described by that theory. Insofar as it contains such statements, it enables us to explain how the entity behaves. Now, assume that frequently, constitutive structures reveal properties of proper parts of our target-entities (as (8) states). Then, for any property *P* that is re-described in terms of its instances' parts and these parts' properties P_1–P_n by a theory, we get several laws that (i) cover and account for the behavior of instances of *P*, and that (ii) play a role for the behavior of objects other than instances of *P* – insofar as they have (proper or improper) parts exemplifying some of the properties P_1–P_n. Thereby, under normal circumstances, in a reduction, we get access to events that happen on the level of parts. Necessarily, this expands the range of possible explanations. The explanatory power of the reduced concept is preserved. On top of that, we get explanatorily relevant information on the object's parts and their properties – as long as reduction tracks mereological dependence. Moreover, by getting access to the constitutive structures, we get better access to *why* the entities we are already familiar with from the reduced theory behave the way they do. This is the second respect in which reduction goes together with an expansion of the range of explanation (this vindicates (9)). It always does. But then, (1) comes for free: When we account for water in terms of H_2O, we explain events amounts of water play a relevant role in by describing water in terms of features of constituents of water-molecules. This, in turn, goes together with the ability to account for events molecules of H_2O as well as their constituents play the relevant role in, such as *forming hydrogen bonds*. Since reduction, in a world where wholes behave the way they do in virtue of their constituents, goes together with a re-description in terms of these parts and their features, the range of explanation is expanded – we can now account for the behavior of the whole in terms of behavior of its constituents as well as for the behavior of its constituents. In general, if a re-description in terms of a constitutive structure gives us more information than the corresponding higher-level property structure does, so that more explanatory links can be established, reduction will necessarily go together with an expansion of the range of explanation. Why isn't it possible that it goes together with a *mere change* in the range of explanation? Because reduction + the transitivity of explanations guarantees that the range of explanation of the reduced level is preserved (we here pick up the topic of theory reduction as direct vs. indirect reduction, see Sect. 7.2). This can easily be shown: Water reduces to H_2O. Freezing reduces to the formation of lattice structures of molecules. Why does water freeze below 0 °C under normal conditions? It does so *because* H_2O molecules form stable lattice structures below 0 °C under normal conditions (i.e. given the right pressure and the fact that the water is not pure). We can, using the resources of our ordinary theory about events of freezing, explain why, say, in Northern Europe, lakes sometimes freeze during winter. They do so *because* water freezes below 0 °C, and during winter, temperature sometimes falls below 0 °C in Northern Europe. If so, by the transitivity of explanation, in Northern Europe, lakes sometimes freeze during winter *because* sums of H_2O molecules form lattice structures when the temperature falls below 0 °C (under normal circumstances). In general:

If F-ness reduces to G-ness, then for every x that is F, it is F because it is G (this follows from the explication). Assume that 'F^*' is a term of (the observational vocabulary of) our reduced theory. Then, by the transitivity of explanation, for every truth of the form 'x is F^* because x is F', that is available in the reduced theory, there is a corresponding truth 'x is F^* because x is G' that is available in the reducing theory. Thus, transitivity plus reduction ensures that the range of explanation is preserved. Given the assumption that constitutive structures are explanatorily more informative, we get the desired result, namely, that a reducing theory has a greater range of explanation.

Here is the connection to (1), that the reducing theory should be better established due to having a wider range of application: Given the definition of reduction proposed here, and given the plausible assumption that the reducing theory contains explanatorily relevant information about instances of the properties which form the constitutive property structure of a reduced kind, as well as the alleged connection between justification of a theory and its range of application, we should expect the reducing theory to be better established than the reduced theory in virtue of having a wider range of application.

8.7.3 Reduction and the Scope of the Reduced Theory

Based on these considerations, we can also account for (7): that the reducing theory explains why the reduced theory worked as well as it did. According to one interpretation, this sort of explanation is directly connected to replacement: Given that the reducing theory replaces the reduced theory, it will show *to which degree* the reduced theory was correct, or how similar it was to the replacing theory regarding the power to adequately predict and explain phenomena (this seems to be what Schaffner (1967, 144), Hooker (2006, 158), and Churchland (1986, 283) had in mind). Once we focus on what is *here* conceived of as genuine reduction, we will get a straightforward answer for why the reduced theory worked as well as it did: It did so because it is *correct*, and, thus, at least part of the evidence for the reducing theory is therefore evidence for the reduced theory. However, there is more to be said about this matter. Reductive explanation does, in another sense, contribute to our understanding of why the reduced theory worked as well as it did. This is directly connected to the idea of reductive explanation. Consider again the following example: Our folk-theory of water enables us to predict when water freezes – this partly depends upon temperature. Once we get access to the behavior of molecules given an amount of energy, we understand why the principles we relied on in our folk-theory of water were adequate. In this sense, once a reduction (as specified above) is achieved, we are in a position to explain, in a straightforward sense, why the reduced theory worked as well as it did. The reduced theory's truths explanatorily depend on truths of the reducing theory. Assume, again, that water reduces to H_2O. Assume that freezing reduces to the formation of lattice structures of molecules. Why does water freeze below 0 °C under normal conditions? It does so *because* H_2O molecules form (stable) lattice structures below 0 °C under normal conditions (i.e. given the right pressure and the fact that the water is not pure, which is, within the range of application of our folk theory about water and freezing events to be expected). Why does our folk theory work as well as it does, i.e. why does it get it right that water freezes below 0 °C under normal conditions, and why would it fail for non-normal conditions? This is so because our folk theory provides the conceptual means to express a truth about freezing of water under normal conditions, but it does not provide the means to account for what these normal conditions are, and what the "mechanisms" are in virtue of which water freezes below 0 °C under normal conditions only. In contrast, and turning back to the previous point about the extension of the range of application, our lower level theory describes water in terms of its constituents. A presentation of water under the relevant constitutive property structure provides the means to account for its properties in virtue of which it does not freeze unless *super*cooled under non-normal conditions. It provides the means to explain events of nucleation, of cluster formation etc. Accounting for water, freezing, and what are supposed to be the "normal conditions" within the lower level science, we can explain why our high level theory got it right to the extent it did.

Note that this is not to say that the reducing theory *directly* explains why the reduced theory worked as well as it did, all by itself. The reducing theory gives the *resources* to account for the entities and events picked out or described by the high-level descriptions by re-describing them in lower-level terms and, thus, under different property structures.

8.8 Conclusion

The explication proposed here does not only shed light on conceptions of reduction in the philosophy of science; it is also more fundamental than its most prominent rivals, namely, explications according to which reduction is a relation that holds primarily between theories. We are, thus, here:

Q2: How can this explication be further motivated?
Th. 7: It is as committal as and more fundamental than rival explications.
...

The explication of reduction in terms of property structures proposed here is more fundamental than notions of theory reduction, it is not more expensive in semantic or metaphysical respects, and it does not, unlike holistic notions of reduction, rely on conditions arrived at by *extension first*-criteria, i.e. by the examination of alleged cases of reductions. These conditions are in need of explanation. This explanation is provided by the explication of reduction offered above (we will turn to the explanation of why reduction goes together with at least some forms of unification in the next chapter). Moreover, the definition proposed here does not face a problem when confronted with ontologically equivalent pairs of reduced and reducing theories.

The explication proposed here does not only meet the job-description; in addition, it seems to yield an explanatorily powerful, unified account of reduction. The range of application of this explication is, however, not limited to the reduction debate, as the next chapter will show.

Bibliography

Churchland, Patricia. 1986. *Neurophilosophy*. Cambridge, MA: MIT Press.

Enç, Berent. 1976. Identity statements and microreductions. *Journal of Philosophy* 73: 285–306.

Hooker, Clifford. 1981. Towards a general theory of reduction. Part I: Historical and scientific setting. Part II: Identity in reduction. Part III: Cross-categorical reduction. *Dialogue* 20: 38–59, 201–236, 496–529.

Hooker, Clifford. 2006. Reduction as a cognitive strategy. In *Paul Churchland*, ed. Brian L. Keeley, 154–174. Cambridge: Cambridge University Press.

Kuipers, Theo. 1982. The reduction of phenomenological to kinetic thermostatistics. *Philosophy of Science* 49: 107–119.

Nagel, Ernest. 1961. *The structure of science. Problems in the logic of explanation*. New York: Harcourt, Brace & World, Inc.

Nagel, Ernest. 1970. Issues in the logic of reductive explanations. In *Mind, science, and history*, ed. Howard Kiefer and Milton Munitz, 117–137. Albany: SUNY Press.

Schaffner, Kenneth. 1967. Approaches to reduction. *Philosophy of Science* 34: 137–147.

Schaffner, Kenneth. 1993. *Discovery and explanation in biology and medicine*. Chicago: The University of Chicago Press.

Sklar, Lawrence. 1967. Types of inter-theoretic reduction. *British Journal for Philosophy of Science* 18: 109–124.

van Riel, Raphael. 2011. Nagelian reduction beyond the Nagel-model. *Philosophy of Science* 78: 353–375.

Chapter 9
Reduction Beyond the Reduction Debate

Reduction reconciles diversity and directionality with strong unity. Diversity is descriptive, or conceptual in nature. Directionality is the directionality of explanatory dependence. Unity is cashed out in terms of identity. The direction of reduction depends upon features of the descriptions under which an object is presented by the expressions flanking the reduction predicate. These features of the descriptions under which an object is presented by the expressions flanking the reduction predicate can be captured by the notion of a property structure. This final chapter contains an outlook on the connection between reduction as conceived of here and the notion of a scientific level, physicalism, various forms of unification, grounding, and intervention. All these topics have been studied extensively; the present chapter thus merely suggests that the so explicated notion can be used to illuminate some aspects of these topics that clearly transcend the boundaries of the various reduction-debates in the philosophy of mind and the philosophy of science.

Section 9.1 introduces two notions of a scientific level and defines two notions of physicalism. Section 9.2 connects the explication of reduction to issues of unification, thereby completing the argument started in the previous chapter, Section 8.7, showing that the model of reduction proposed here is more fundamental than holistic notions of reduction. Section 9.3 suggests that there is a tight connection between reduction and grounding, and Sect. 9.4 characterizes one pragmatic dimension of reductive explanation: it gives access to specific intended interventions.

9.1 Reduction, Scientific Levels, and Physicalism

The explication of reduction proposed above sheds light onto the notion of a scientific level in a reductive hierarchy. Levels turn out to be conceptual levels, or levels of kinds of property structures. The physical level is distinct from the mental level in a reductive hierarchy because the former employs physical property structures, i.e. structures that present us with an object as physical, whereas the latter employs mental property structures that present us with an object as mental. This is

R. van Riel, *The Concept of Reduction*, Philosophical Studies Series 121,
DOI 10.1007/978-3-319-04162-9_9,

the sort of levels a reductionist is committed to, not more and not less. We thus deliver some of the details needed to understand generic reduction-statements, such as 'mental properties reduce to physical properties' (see Sect. 5.6.2). To see how the notion of reduction proposed above connects to level-talk, consider the question of what physicalism consists in.

9.1.1 Physicalism

What would the reductive physicalist want her notion of physicalism to achieve? It should (i) give an idea of how objects at different scientific levels differ, and (ii) give an idea of what the objects that constitute our world basically are. The framework proposed here suggests a disambiguation: In one sense, an object belongs to a scientific level insofar as it is presented under property-structures distinctive of that level. An object belongs to a scientific level in another sense insofar as it *can* be presented under property structures distinctive of that level. Physicalism is, then, the doctrine that any object is physical in the second sense, and that, in principle, everything reduces to physics. The property structures distinctive of physics are, according to physicalism, the most transparent with respect to the nature of objects that make up the actual world; based on these, we can explain all the rest.

To motivate this interpretation of two senses of physicalism, let me introduce a puzzle that is similar in spirit to the one that guided the discussion in Chap. 3. Within reductive hierarchies, we can distinguish different levels. The terms 'biological level', 'chemical level', 'psychological level' and so forth can be used to refer to levels within a reductive hierarchy (neglect sciences that clearly span levels for the moment). Moreover, within a reductive hierarchy, the biological level is clearly not the physical level. Similarly, *water, pain* and even *C-fiber stimulation* do not belong to the physical level in any reductive hierarchy they occur in. At the same time, one might want to hold that if everything is physical, then everything belongs to the physical level. This seems to be knowable *a priori*; or it seems to be an obvious condition on any appropriate explication of the thesis that everything is physical as well as of the notion of a physical level. Assume thirdly that physicalism might be true even in worlds with reductive hierarchies comprising more than the physical level. This seems equally plausible: Reductionism (based on the notion of reduction explicated above) should be compatible with physicalism. This requires there to be at least two levels – a reducing one and a reduced one. From these three assumptions, we get the following picture:

There is a possible reductive hierarchy in a physicalist world which is such that for some x, x belongs to the, say, biological level of that hierarchy and, thus, not to the physical level of that hierarchy and is, thus, biological and not physical (since in a reductive hierarchy, it cannot be both at the same time). At the same time, since physicalism is true, x is physical. So, doesn't it belong to the physical level? This pretty much looks like a contradiction.

This puzzle is of heuristic value, and the solution is easy to sketch: There are at least two different notions of *being physical* (or, more generally, of belonging to level *F*). The sense in which an object is not physical although it is identical to some object situated at the physical level, is this: It is not physical *under some property structure,* or *under some meaning,* but *physical under another.* So, being physical depends, in one sense, on being presented under a specific meaning (and being presented under a specific property structure the meaning gives access to). Thus, *being physical* would be a two place relation – an entity would be physical with respect to some property structures under which this entity is presented, and not with respect to other property structures determining this entity.[1] On this interpretation, predicates such as '_is physical', '_is biological' and the like are hyper-intensional: substitution of co-referential expressions may alter the truth-value.

The second notion is a notion of *being physical* (or *being an F*-entity, or *belonging to scientific level F*) that is independent of how an object is presented on a given occasion. I will come back to the distinction between two notions of being physical in a moment. Let us first consider what features of a property structure might determine which level it is associated with. I shall just outline a general framework, without trying to settle this issue finally, and I will dwell upon several examples.

9.1.2 Scientific Levels

The idea is simple: different reductive levels are associated with a coherent pool of properties that form complex property structures at these levels. Any such coherent pool of properties forms the set of basic constituents of the property structures at the corresponding level. Thus, to which level a property structure belongs, or whether or not it spans levels, depends upon which properties are its basic constituents. Consider the following quote from Cussins:

> We perceive the world as beautiful or ugly, sweet or salty, happy or sad, brave or cowardly, intelligent or stupid; yet none of these properties figure in the world of the physical science. (Cussins 1992, 179)

If we replace 'in the world of' by 'as basic in the property structures of', we get an idea of how different properties might play different roles at different levels. Perceptual properties, or qualia-properties are basic in our everyday representations of the world. They are not basic in our physical representations of the world. Nevertheless, they may be physical in nature and, thus, be representable by a

---

[1]Note that this is in line with Fodor's conception of a kind, that is (i) tied to the notion of a law, and (ii) hence, at least intensional: Fodor suggests that 'it is a law that_' is not truth-functional (Fodor 1974, 109). This also seems to correspond to an idea suggested by Hempel in his (1969), where he argues that a purely ontological characterization of what it is to be physical, or biological, or the like, might fail.

property structure the basic constituents of which are properties that are basic in physical representations. Let me point to the fact that this assumption perfectly matches several ways of conceiving of levels which are implicit in the debate, thereby giving examples of what one might expect to be a *coherent pool of properties*:

(a) Early behaviorists argued that psychological terminology is analyzable or in some other sense conceptually reducible to behavioral terminology. If this is true, behavioral properties figure as basic constituents of property structures of psychology. In this respect, psychology would have to be distinguished from other sciences, in which other properties play a similar role.
(b) Modern functionalists conceive of the meaning of psychological terminology as being given by relations to other psychological terms and to behavioral terms. Again, basic constituents of such property structures will be behavioral properties (and some concepts of causation and an internal state). The difference to early behaviorists is not that we change the set of basic properties, but rather that we do not assume that there is a straightforward mapping function from psychological terms onto input–output relations without invoking sensitivity to different internal states which might influence each other. Thus, again, there is a coherent set of properties, which figure as basic in property structures employed by psychology according to the functionalist.
(c) Assume that qualia concepts cannot be conceptually reduced to behavioral concepts (see, for example, Papineau 2007). Then, there are genuine qualia properties that figure as basic in property structures under which qualia-terminology presents us with internal states. Let this level be the *phenomenal level*. Again, there is a coherent set of properties (namely: qualitative or phenomenal properties), which figure as basic in property structures characteristic of the phenomenal level. This is obviously not to say that these properties are not reducible. Rather, a description's feature of *being phenomenal* is described via the set of properties that are presented as basic in property structures of that level. Assuming that phenomenal properties are physical, a physical description of phenomenal states would present these properties under property structures that do not involve phenomenal properties *as basic constituents*; rather, they would determine phenomenal properties by giving their physical architecture.

I do not claim that these specific conceptions of scientific levels are correct. I merely claim that they perfectly match the assumption that there is one coherent set of properties which function as the basic constituents of the relevant property structures. But what does account for the differences between these pools of properties? This is a topic for another book. One might speculate that the relevant pool of properties depends upon the set of experimental or observational techniques that are applied investigating a specific set of phenomena – that would explain the occurrence of sciences that "branch" classical scientific levels as well as differences between the levels of the special sciences. Biochemistry consists in the study of chemical aspects of phenomena classically studied in biology, obviously with the application of techniques common in chemistry. Thus, intuitively, every science comes with its basic kinds. Clearly, these basic kinds may gain special attention within a science and are, maybe, the science's primary target of investigation (this is, basically, Causey's point claiming that there is a sub-set of the relevant ontology a theory or science is *primarily* concerned with, although it may account for more phenomena (Causey 1972, 1977)). However, they do not form the set of kinds a science is concerned with *only*; rather, they figure as basic in property structures associated with this science. The kind *reproduction* or *being an organism*, or related properties, figure among the basic properties of some parts of biology, and it is this fact that makes these property structures the structures that are associated with biology. Behavioral properties and properties concerning "internal" relations form the basic properties of folk psychology, if functionalism is correct. Briefly: to characterize what makes a science belong to the reductive level it belongs to is to characterize it in terms of the properties that figure as basic in the property structures this science gives access to. In this sense, we can distinguish kinds of property structures. We can thus make sense of generic reduction statements (see Sect. 5.6.2). Let me repeat: *Mental* properties reduce to *physical* properties; For every property x, if x is determined by some mental property structure z, then there is a property y, a physical property structure z^*, such that x, when presented under z reduces to y, when presented under z^*.

Prima facie, we have reason to assume that the difference between levels stems from a difference in properties that function as basic constituents in the property structures characteristic of this level. In this sense, being an F-entity (with 'F' standing for some term like 'biological', 'chemical' or the like) depends upon which basic properties form the property structure under which the object is presented. In the light of this, we can now turn back to the two notions of a scientific level and easily solve the "puzzle":

The Puzzle's Solution: Two Ways of Being Physical (at least)

Being an F-entity, i.e. being a physical, biological, neural ... entity, in one sense depends upon being presented by elements of the set of properties characteristic of the F–level:

(Def. Being an F-entity) *An entity a is an F-entity iff the meaning of 'a' gives access to a property structure of level F.*

This sense of being situated at a level is tied to characterizing something as being biological, physical and the like within *reductive hierarchies*. Thus, '_is physical' is, in this sense, hyper-intensional: It yields a truth only if the referent of the argument is conceptualized appropriately *by* the argument. Now, consider the second sense in which something might be *physical*. According to this sense of being physical, biological and so forth, we get the following criterion:

(Condition – Being an F-entity)* *An entity is an F-entity independent of how it is presented.*

This captures notions of being physical that rely on purely ontological descriptions of being physical. It is pertinent in explications of the notion of being physical which refer to physics (Smart 1978; Braddon-Mitchell and Jackson 1996; Chalmers 1996), and those which describe being physical via paradigmatic physical objects and their properties (Meehl and Sellars 1956) (thus, the "second sense" of 'being physical' is, maybe, a family of different notions of being physical, which are tied together by taking things to be physical in some sense which is independent of how these things are presented). However, we can, relying on the hyper-intensional sense of belonging to a level, as described by *(Being an F-entity)*, give a criterion for belonging to a level in the second sense:

(Def. Being an F-entity)* *An entity is an F-entity iff it is determined by a property structure the basic constituents are genuine to the level F.*

Accordingly, an entity would be physical iff it is determined by a property structure the basic constituents of which are typical of the physical level (neglecting logical constituents). That is: It is physical if, in principle, it could be presented under a property structure characteristic of the physical level. And it is mental if it, say, makes for a phenomenal or behavioral difference.

This picture enables us to *explain what identifying some high-level property with a physical property consists in.* Some property P (= property P*) is at a higher level in a reductive hierarchy than P* iff the property structure associated with 'P' belongs to some higher level within that reductive hierarchy than the property structure associated with 'P*'. However, P (and, thus P*) might be a physical property. There is, then, no puzzle of how some property can be mental and, thus, in one sense non-physical, and at the same time physical. This also bears upon the relation between ontological and reductive hierarchies: Ontological hierarchies, whatever these are, do not perfectly match reductive ones. The former are insensitive to how its objects are presented, the latter are not.

This distinction is of utmost importance when we try to distinguish between purely ontological issues on the one hand (concerning, say, physicalism, monism and so forth) and issues that are sensitive to ways of presenting or describing an object (like issues of reduction). The concept of reduction is, thus, not the only notion in the field that is sensitive to meanings; in one of their uses, predicates such as '_is a physical kind' and '_is a psychological entity' are used to form sentences the truth of which is sensitive to semantic features other than designation or reference of their arguments. Given that the model of property structures captures the relevant aspect of meaning, levels are, at least partly, individuated by property structures.[2] Levels present things in different ways because they present things

[2]Here is an application of this idea. Under the assumption that property structures fully capture the cognitive role of meaning, so that there is no property structure associated with two cognitive roles, this can be used to describe one condition on the truth of *a priori* physicalism. A priori physicalism states that *from a statement of all physical facts, all truths about the mental are knowable on a priori grounds.* (Beckermann 2009, 162) For a priori physicalism to be true, there should be some a priori path from the property structures of the physical level to property structures at higher levels (if property structures are the primary targets of a priori operations). Then, a priori physicalism is

under different properties. Theories or other representational items at distinct levels may be concerned with the same set of entities and kinds and yet be distinct.

This suggestion also points to how reduction relates to various aspects of *unification*. In general, this strategy of tying qualifications such as '_is physical', or '_is mental' to presentation under sorts of property structures enables us to account for descriptive *pluralism* in an ontologically *homogeneous* world, or: to maintain ontological *unity* while allowing for descriptive *pluralism*. The hints towards unification in the reduction literature are rather vague. The next section argues that given the explication offered here, a number of different sorts of unification are to be expected for cases of theory reductions (we thus not only provide further reason for the claim that the explication proposed here is fruitful; at the same time, we resume the issue of Chap. 8: the argument that the account proposed here is more fundamental than holistic accounts).

9.2 Reduction and Unification

The topic of unification in science forms a debate in its own right, especially since it has been described as being closely tied to explanation in general (Friedman 1974; Kitcher 1981, 1989). Scientific unity is discussed under several distinct foci, and one might easily get the impression that the debate is even more complex than the debate on reduction.[3] Here, we cannot give anything that comes even close to a full-blown account of how reduction relates to scientific unity in all these respects. Thus, I shall just point to how it relates to ontological, epistemological and explanatory unification. The three remaining conditions for directionality in holistic approaches (2–4) to reduction can thereby be shown to depend on reduction as explicated here.

---

true *only if a complete description of the world under physical property structures conceptually implies, or gives a priori access to, the set of truths about mental facts presented under mental property structures.*

It has been assumed that what makes a given property structure belong to the physical level within a reductive hierarchy is that it is constituted by elements of a specific set of properties which are presented as basic. Thus, a physical description of everything would be a description that gives access to everything *under these basic properties (and logical operations on these basic properties) only*. Now, higher-level descriptions give us entities under *their* specific property structures. In this specific sense, there has to be an a priori route from the physical level to higher-levels, if a priori physicalism is true.

[3]Critics of the unity-program often neglect the fact that this program did not consist in attempts to actually carry out unifications. Recall the Kemeny and Oppenheim-remark on a definition of reduction: such a definition would become "hopelessly complex" (1956, 13), if we were to capture all the relevant aspects of actual or possible reductions. In these and similar writings, it is quite common to appeal to an ideal demand rather than arguing that such general reductions could actually be carried out. At least, this point did not receive the attention it deserves. Although unification might not be a goal one should actually try to accomplish in any important sense, it might nevertheless be more than a regulative idea– it captures one aspect of the idea that some representation is more appropriate, or more basic than another.

9.2.1 Reduction and Ontological Unification

The notion of reduction, and the two sorts of notions of belonging to a scientific level F (a physical level, a biological level and the like) described in Sect. 9.1 give a precise idea of how to maintain *ontological unity* while enabling us to account for *descriptive pluralism* in a very specific sense of these terms. Ontological unity for a set of entities (kinds, states, objects and so forth) holds iff every element of this set can be described under one coherent set of property structures. Descriptive pluralism for a set of entities holds iff every element of this set can be described under at least two distinct coherent sets of property structures. The answer to the question of how this is possible is straightforward: Descriptive pluralism is required by reduction – each reductive level comes with its own property structures under which entities are presented at that level. With regard to their descriptions on given occasions, entities may be physical, biological and so forth and at the same time belong to one specific ontological category, like being physical. Different ways of being an F-entity allow for there to be different descriptions which give us just one (ontologically construed, or description-independent) kind of entities.

Given that ontological unity and descriptive pluralism hold at one world, *reduction* is the relation that ties together descriptive pluralism and ontological unity. In virtue of reducing to the lowest descriptive level (in virtue of being presentable under the property structures characteristic of this level), entities are ontologically situated at this level. F-entities (say, biological entities) reduce to F^*-entities (say, chemical entities) iff they are identical to F^*-entities and the F-descriptions or -property structures are adequately tied to F^*-descriptions or -property structures. In virtue of being presented under different property structures at different levels, the reduction task is an interesting one: We have to find the relevant cross-level identities. Once this is done, we may hope for reductive explanation and, thus for *ontological unification*. Picking up the topic of Chap. 8: Again, one of the features invoked in holistic accounts of the directionality of reduction, namely, ontological unification, is to be expected based on, and explained by the framework proposed here. Ontological unification occurs when we acquire knowledge about identities of objects that seemed to be situated at different ontological levels, which turn out to be mere reductive, and, hence, descriptive levels. The idea that reduction is tied to unification as a procedure of showing that there is unity where there seemingly was not has important *epistemological* implications.

9.2.2 Reduction and Epistemological Unification

Let us now turn to the epistemological value of the procedure of unification in science. *Carrying out a reduction goes together with epistemological unification.* The explication given above enables us to see *how it is possible that there are true sciences (or, more liberally: appropriate descriptions and terms) that are seemingly unconnected, or are situated at different levels, although they are compatible with strict monism.* This is in need of further clarification. Epistemological unification, I suggest, consists in showing that what was presented by two different sorts of property structures (or meanings) can be *presented* by one. This captures the interpretation given for Nagel's characterization of reduction as an *assimilation of traits*.

Assume that two theories T_R and T_R* both reduce to some theory T_B. (The same point could be made using just one reducing theory. However, using two makes the conceptual and the ontological aspect more transparent.) Assume, moreover, that T_R and T_R* seemed 'unconnected' (Nagel 1961, 359). One criterion for seeming unconnected is being such that there is no (relevant) a priori or known metaphysical connection between statements of T_R and T_R*. If there were no a priori connections, then, according to the model outlined above, both theories were associated with relevantly different property structures. Now, introducing bridge-principles which state identities and subsuming what was described by T_R and T_R* under *one common conceptual framework*, (presenting their respective fields under property structures which use the same resources of basic properties), namely that of T_B, is clearly some sort of *unification of the conceptual resources*. What seemed unconnected before is now shown to be connected – to be representable under one common conceptual framework. Thereby, the conceptual aspect of unification is captured; since knowledge is mediated by conceptual presentation, we could describe this sort of unification as epistemological: One unified body of knowledge (in terms of T_B) captures what involved a body of knowledge in terms of T_R and a body of knowledge in terms of T_R* before.

This conceptual or epistemological aspect goes together with the ontological aspect of unification described in the previous section. By the reduction of $\mathrm{T_R}$ and $\mathrm{T_R}$* to $\mathrm{T_B}$ it is shown that the difference in *meaning* of the languages of $\mathrm{T_R}$ and $\mathrm{T_R}$* (and $\mathrm{T_B}$) bears no relevant ontological implications. There just is one ontological sort of entities, although there are different sorts of entities with respect to how they are presented. Thus, the intuition that reduction is tied to unification, or that reductions go together with unification, can be given a precise explication. Two aspects of unification, namely, a conceptual aspect, which bases an epistemological, and an ontological aspect can thereby easily be explained using the model outlined here. Again, a holistic criterion turns out to be derivative on the explication proposed above. In addition, reduction amounts to what one may want to label *explanatory* unification.

9.2.3 Reduction and Explanatory Unification

The explanatory priority of the property structures given in the *explanans* of a reductive explanation corresponds to the degree to which these properties are responsible for how the reduced object behaves. One interesting form of explanatory unification can easily be characterized as follows, building on an idea already hinted at in Sect. 8.7.2. We already know that in a theory reduction, the range of explanation is expanded. Let me repeat. Water reduces to H_2O. Assume, again, that freezing reduces to the formation of lattice structures of molecules. Why does water freeze below 0 °C under normal conditions? It does so *because* H_2O molecules form (stable) lattice structures below 0 °C under normal conditions. Given our every day theory of freezing events, we can explain why, say, in Northern Europe, lakes sometimes freeze during winter. They do so *because* water freezes below 0 °C, and during winter, temperature sometimes falls below 0 °C in Northern Europe. If so, by the transitivity of explanation, in Northern Europe, lakes sometimes freeze during winter *because* sums of H_2O molecules form lattice structures when the temperature falls below 0 °C (under normal circumstances). This follows from the explication of reduction and the transitivity of explanation:

If F-ness reduces to G-ness, then for every x that is F, x is F *because* x is G. Assume that 'F^*' is a term of our reduced theory. Then, by the transitivity of explanation, for every truth of the form 'x is F^* because x is F', there is a corresponding truth 'x is F^* because x is G'. Thus, transitivity plus reduction ensures that the range of explanation is preserved. Reduction + transitivity + the idea that a constitutive property structure is always more informative as to the nature of the object (for example, by tracking mereological dependence) than the higher-level property structure ensures that the range of explanation is expanded. Now, applying the notion of a scientific level, we arrive at explanatory unification: If property structures at a level come in a uniform format, then the idea that reduction guarantees *explanatory uniformity* comes for free. By a reduction, we get a greater range of our conceptually coherent *explanatory resources*. One coherent set of property structures enables us to explain what we used to explain using a different set of property structures. In this sense, we get explanatory unification, which is, somewhat like epistemological unification, a side effect of conceptual unification. What we gain by reductions (in this sense) is *transparency* with respect to the contribution of properties to what we want to explain. Reduction goes together with conceptual (and, hence, explanatory and epistemological) unification as well as with ontological unification. Hence, another holistic criterion turns out to be derivative on the explication proposed in the first part of this book.

As a concluding remark on unification, it should be noted that a great deal of the debate on unification is concerned with the scientific *value* of unification (an idea which has been attacked by Darden and Maull (1977), Cartwright (1983, 1989), and Dupré (1993)). Let me just point to the fact that the interpretation of the connection between reduction and unification I proposed by no means goes together with the idea that unification is the *only* goal of science. However, it *shows* that it is at least one goal (which might even be incompatible with others).

Recall the distinctions suggested in the introduction concerning the connection between realist/anti-realist and conservative/eliminative versions of reductionism, that correspond to different interpretations of the alleged truth or relevance of high-level theories or sciences (see Table 9.1).

Table 9.1 Monisms and reductionisms

Monisms and reductionisms	Anti-realism	Realism
Eliminativism	Monism-1: Eliminativism/replacementism	Monism-2: Unificationism
Conservatism	Monism-3: Epistemic/pragmatic non-reductivism	Monism-4: Conservative reductionism

In the introduction, I mentioned that there is no conservatism or eliminativism (building on normative constraints on reductions) *tout court*. The book deals with realist versions of reduction, i.e. versions that assume that in principle, high-level sciences or theories or descriptions may be true, or that the entities purportedly referred to by these theories and descriptions possibly exist. For these cases, we should explicitly distinguish different respects in which eliminations might be appropriate or misguided. Elimination (in the sense of getting rid of a theory or a way of talking or conceptualizing) is, in one specific respect, a goal worth aiming at, because it will force us to reject less transparent theories and replace them by more transparent ones (or move to reductive levels the property structures of which are explanatorily prior to the property structures associated with higher levels). Once we give primacy to what is explanatorily prior, we clearly gain knowledge about the structure of our world, if it is monistic. As pointed out above, once we get a reduction, we get unification, and as long as we regard unification as a goal in itself, this might give further reason to believe that reductions are relevant and worth aiming at.

On the other hand, there are many respects in which elimination is *not* a goal worth aiming at. Or, as Elliot Sober put it: "The reductionist claim that lower-level explanations are *always* better and the antireductionist claim that they are *always* worse are both mistaken." (Sober 1999, 560, italics in the original). One problem is, clearly, that once descriptions become too complex, we would be utterly lost. In this respect, some reductions might not be worth aiming at. Accordingly, (and again) we should distinguish different respects in which levels of description are appropriate. Some (possible) reductions would supposedly contradict criteria of pragmatic appropriateness (see also van Gulick 1992; Friedman 1982, 17; Wimsatt 1976).

One cognitive role of high-level descriptions (which shows that they are indispensable in one respect) can nicely be illustrated using a quote from Lennon (1992):

> The distinct physical groundings of a given intentional kind will be *shapeless* from the viewpoint of physical theory. [...] Our classifications into intentional kinds provides a pattern of conceptualization which yields empirical generalizations and conditional dependencies. These patterns of dependency are at the heart of causal explanation, and could not be captured without the employment of intentional notions. (Lennon 1992, 226)

What sort of modality is involved in 'could not be captured without the employment of intentional notions' (which are not to be confused with kinds, as this passage suggests)? A modest interpretation would suggest that it is *us* who could not capture the relevant dependencies without using these (or similar) notions, due to our cognitive shape. This interpretation is compatible with reduction as the notion is spelled out here: For us the complex physical descriptions of high-level kinds might look shapeless. But this is not necessarily so. If we were able to grasp the property structures revealed by the relevant low-level descriptions, the apparent reason to assume that these descriptions look shapeless might disappear. Therefore, we should not assume that if one representational item is more appropriate with respect to, say, practical or didactical purposes, it is also a more appropriate device in the sense that it represents the relevant domain in a more transparent way. What reduces in virtue of its representational structure does not necessarily "reduce" when questions of accessibility are addressed. More precisely, possibly implicit aspects of *normativity* in the notion of reduction are limited and context-sensitive. There isn't just one goal. Gaining transparency and unification form one goal, systematization (Kemeny and Oppenheim 1956; Kitcher 1981, 1989) is another. In principle, it would be great to achieve these goals all at once. But due to our limited cognitive capacities, this hope seems to be futile. So, we should be flexible in our use of notions of *eliminative* or *conservative reduction* (in the pragmatic sense): Reduction is clearly not the *only* goal of science. Let us now turn to another issue, that of connecting reduction to what has recently been discussed under the title 'grounding' in metaphysics.

9.3 Reduction and Grounding

This section briefly connects the book's topic to a discussion that emerged in a completely different area of philosophy – a topic closely related to ontological dependence. The goal of this section is modest; it argues that, under a very general conception of what theories of grounding are after, reduction is a promising candidate for counting as a *variant* of grounding.

Classical accounts of non-causal and, usually, non-conceptual dependence relations have tried to come up with explications in terms of metaphysical modality (see, for example, Simons 1987). Recently, this view has been attacked by, for example, Correia (2005), and Fine (2005).[4] There are two sorts of criticism of this approach that are relevant for the present discussion. One criticism concerns what Fine dubbed "modal-mania" in metaphysics (2005, 9). Roughly, it is argued that modal notions cannot be used to define a relevantly *directional* relation, or, in the syntactic sense, a directional expression. Thus, focus shifted towards explanatory concepts. And this is where the second criticism enters: Fine (2001) argued that 'because' and similar expressions do not express a *relation*. This latter point will be ignored for the present discussion, although it should be noted that if it is correct, then the *reduction*-relation as described here is not that of explanatory dependence. The reduction relation is

[4]For an overview, see Correia (2008).

then conceptually tied to non-relational explanatory concepts.[5] Moreover, I will not question the first criticism. Rather, I want to briefly address the question of how reduction relates to grounding, if grounding is not to be cashed out in counterfactual terminology.

In a nutshell, it will be suggested that one promising job-description of grounding or grounding-relations is this: grounding reconciles *diversity* and *directionality* with *committal unity*. Depending on what we take to be the primary relata of grounding, grounding is *committal* with respect to the grounded and the grounding entity; that is: either with respect to the truth of what is grounded and what grounds or to the existence of what is grounded and what grounds. Put differently, grounding, as understood here, is distinct from elimination. If *that p* is grounded in *that q*, (and it is true *that q*) then *that p* is true. Insofar as strong unity is a form of *committal* unity, reduction, as explicated here, is a special case of grounding.[6]

Jenkins (2011) has proposed a very general argument for the claim that reduction is a special case of grounding. She suggests that we might *want*, in principle, our notion of grounding to be neutral with respect to the question of whether or not if *a* is grounded in *b*, *b* and *a* are identical.[7] This section is supposed to further motivate this idea, reflecting on the relation between the two job-descriptions for 'ground' and 'reduction'. To begin with, recall the *prima facie* plausible worry that immediately arises, which is based on an idea already hinted at (Chap. 3):

If *a* reduces to *b*, then $a = b$. Therefore, reduction has the formal properties of identity: reflexivity, symmetry, and transitivity. Following this line of thought (*cf.* Trogdon 2013, Section 6), one might go on stating that grounding is *irreflexive*. Therefore, reduction is not a variant of grounding, even if it does not exhibit the structural features of identity (as has been argued above). Reduction is not irreflexive either; so, if grounding is, then reduction is not a variant of the grounding relation. But is grounding irreflexive? Obviously, we should assume that if an instance of '*a* is grounded in *b*' expresses a truth, then the corresponding instance of '*b* is not grounded in *a*' expresses a truth as well. However, this does not, without further ado, imply that the grounding relation is irreflexive; Jenkins (2011), for example, argues that it is not irreflexive, reflecting on cases of reduction. How can we decide? Isn't this a matter of convention, or, again, stipulation? To assess the question of whether or not reduction is a form of grounding, we should first get clear about what this question actually consists in.

---

[5]This raises further subtle issues: If expressions such as 'because' do not express relations, one may ask whether or not expressions such as 'grounds' or 'ontologically depends upon' are themselves appropriately explicated in terms of relational concepts (*cf.* Fine 2001). Reduction itself may be inadequately described in relational terminology; however, a decision on that issue would take us too far from our present target. If explanatory dependence is not relational, then the explication proposed above should be rephrased accordingly.

[6]Note that the use of 'reduction' deviates from the way it is introduced by Fine in his 2001.

[7]Her discussion of questions regarding asymmetry and reflexivity arrive at conclusions similar to the ones suggested above.

9.3.1 Reduction as Grounding – The First Argument

'_grounds_' and '_reduces to_' are more or less technical expressions a characterization of which should take the form of an explication (Chap. 2). Correspondingly, the question of whether or not a technical concept *F* is a variant of a technical concept *G*, could be rephrased as follows: According to the best explications of *F* and *G*, is *F* a variant of *G*? Obviously, an explication of the concept of *grounding* cannot be given here. We thus cannot give a straightforward answer to the question of how reduction relates to grounding. Two alternatives suggest themselves: We could go through the different candidate explications for *grounding* and check for each of them whether or not it covers reduction as well, and then give a conditional answer: given the appropriateness of the account, reduction is/is not a case of grounding. One problem with this procedure is that it does not provide the resources for an answer to the following question: *Should* grounding be regarded as covering cases of reduction?[8]

Any fruitful characterization of grounding should, like any other explication, comply with a job-description. The question of whether or not reduction is a form of grounding can then be construed as a question concerning the respective concepts' job-descriptions: Is fitting the job-description for reduction a way of fitting the job-description for grounding? If the above job-description is correct, reduction is a form of grounding: Whenever we have reconciliation of directionality and diversity with *strong* unity, we have reconciliation of directionality and diversity with *committal* unity. Thus, we now have to motivate the idea that the proposed job-description for grounding forms at least one appropriate way to characterize the idea underlying grounding talk.

It will prove useful to think of the predicate '_grounds_' and its variants as expressions whose definitions have to correspond to a number of uses of *other* expressions that are taken from samples of ordinary, scientific, and philosophical discourse, which, under an appropriate interpretation, inform the job-description. Schaffer suggests the following list: '(i) the entity and its singleton, (ii) the Swiss cheese and its holes, (iii) natural features and moral features, (iv) sparse properties and abundant properties, and (v) truthmakers and truths.' (Schaffer 2009, 375)

These samples comply with the job description already mentioned, in slogan form: *Grounding reconciles diversity and directionality with committal unity* (that it is an atemporal or synchronic relation should go without saying). This rough

[8]To answer this question, we could try to check explications of grounding on the market; however, grounding is often regarded as primitive (Schaffer 2009; Fine 2001). But even if there were "reductive" characterization, like in the case of reduction, one should not treat the characterizations as stipulations; rather, they are *proposals*. One might even feel tempted to revise a given proposal in the light of the fact that they cover, or do not cover a notion of reduction (this seems to be Jenkin's move). Thus, showing that reduction is covered by alleged characterizations of grounding is not enough to account for the relation between reduction, as construed here, and the notion of grounding the candidate characterizations are after. The alternative strategy is maybe less straightforward, but it seems to perfectly suit the present purpose.

description can be further motivated reflecting on further examples; the expressions 'grounds' and 'is grounded in' can be employed to capture metaphysical relations that are expressed by or correspond to metaphysical explanations, and their use is informed by the use of other bits of philosophical jargon, such as 'constitutes', or 'brings about', and more technical expressions such '_is an element of_', and, maybe, '_is an epiphenomenon of_'. Here are a few examples:

(i) The field exists *because* the individual plants exist.
(ii) My headache occurs *because* a number of physiological events occurred.
(iii) This statue is beautiful *because* it has specific physical properties.
(iv) The heart pumps blood *by* regular muscle-contraction.
(v) This statue is *constituted by* a chunk of matter.
(vi) My mental processes *do not exist over and above* neural processes.
(vii) The mayor's announcing you husband and wife *brings it about* that you become husband and wife.
(viii) My current mental state *is an epiphenomenon* (given a non-causal interpretation of 'epiphenomenon') *of* my current brain-state.

I submit that none of the statements in the group (i)–(iv) expresses a truth unless the explanantia and the explananda are true and/or the entities the explananda are about exist (one may describe this as follows: theories of grounding buy into the factivity of explanations); moreover, none of the members of the group (v)–(viii) is true unless the relevant entities exist. Thus, if grounding is supposed to capture these fragments of discourse, it is *committal*. Moreover, we get unity by directionality: The cases suggest that there is a fundamental level that fixes the rest. In this respect, (i)–(viii) are similar to the examples on Schaffer's list. To the extent that this already gives a very general notion of grounding, reduction, as defined above, is a particular form of grounding. It is a form of non-causal, non-conceptual dependence, that is possibly cashed out in terms of explanation and it is committal in that it does not rely on a purely eliminative stance towards the grounded/reduced objects or propositions. Grounding reconciles diversity and directionality with committal unity. Reduction does so, too, in a specific way: Diversity is conceptual, and unity comes in the form of identity. If these job-descriptions are appropriate, we have a very general reason to assume that reduction is a case of grounding. Jenkins proposes an additional argument for the thesis that metaphysical dependence or grounding is (best construed as being) neutral with respect to identity.

9.3.2 Reduction as Grounding – The Second Argument

Jenkins' argument seems to rely on strategic considerations (Jenkins 2011, 269f. – here adapted to the present list of examples): A reductive interpretation of examples similar to those on the list discussed in the previous section seems at least conceivable. If so, and if we want our notion of dependence or grounding to cover these cases, we would be well advised to describe dependence or grounding as being neutral with respect to reduction. Thus, one may *want* to describe grounding in a neutral way, so that by embracing a grounding claim, one is not committed to the non-identity of the *relata*. After all, the cases decide whether or not we are faced with identity, so the conceptual apparatus we employ to approach these cases should remain neutral in this respect (Jenkins 2011, 270). But why not adopt a different (terminological) strategy? Why not employ 'reduction' and 'dependence' or 'grounding' in competing ways, so that reduction and grounding turn out to be mutually exclusive? Then, instead of phrasing the issue in terms of whether or not a particular case of grounding is a case of reduction, we would have to assess the question of whether or not a given case is a case of grounding *or* reduction. Apparently, we do not increase or decrease the expressive power of our theoretical apparatus by opting for this alternative strategy. As a thesis about how we should construct our conceptual apparatus, the point is primarily terminological in nature. But terminological decisions may be strategically relevant. In the present case, having a notion that captures the similarities may form a strategic advantage – we do not lose sight of the possibility that we may be convinced that some x is grounded in y (because the case bears some signature features of grounding), and still, x may turn out to reduce to y, without this having any relevant effect on our conviction that x is grounded in y. *Prima facie*, reductive interpretations are conceivable for various candidates of grounding claims: On some accounts, the field will turn out to be identical to (the sum of) the individual plants (i), just like my headache will turn out to be identical to the relevant physiological events (ii). The latter has been offered as a paradigm case for identity based reduction (Nagel 1961, 366). Cases similar to (i) seem to fit into a reductionist picture as well (though some may hesitate to assume that the field is nothing but the individual plants): A given amount of water reduces to the (sum of) H_2O molecules, which constitute it, and it exists *because* the latter exist. (iii) seems to form a paradigm case for reduction (from the reductionist's perspective). Similarly, as has been extensively argued in Chap. 4, 'by' explanations can be given a reductive interpretation: A heart's pumping blood *might be identical to* rhythmic muscle-contraction in an appropriate environment. And at least *prima facie*, the way the term 'constitution' is employed may have instances where *what is constituted is identical to what constitutes it*: Early type-identity theorists took (complete) constitution-relations to *be* identity-relations (Place 1956, 1960; Smart 1959). According to this interpretation, our mental life is constituted by part of our physical life; at the same time, it just *is* this part of our physical life. Similarly, it goes without saying that the idea that mental processes *do not exist over and above* (and are, therefore, grounded in) neural processes has a straightforward reductionist interpretation. There may be uses in which 'brings about' cannot be coherently interpreted reductively, like in the truth-maker case alluded to above. But why shouldn't a reductionist about phenomenal properties happily admit that pains are *brought about by* (and are, hence, grounded in) C-fiber firings, although pains *are nothing but* C-fiber firings? What about more technical philosophical conceptions, such as *being an epiphenomenon of*, various forms of *dependence* or *being an element of*? Reductive interpretations of epiphenomena, sets and their elements as well as particularized properties and their bearers seem rather counter-intuitive. For epiphenomenalism, which is introduced as an anti-reductionist position, such an interpretation cannot be given without changing the subject. Some nominalists, however, famously argued that sets can be reduced to non-abstract objects (as, for example, sketched in Lewis 1970); should we, then, give a definition of grounding which is such that no nominalist can consistently hold that Sokrates *grounds* {Sokrates}? Again, it might be a good idea to remain neutral on whether or not our use of 'grounding', when used to capture a relation between a set and its elements, commits us to the assumption that a set is distinct from its elements. In general, it seems to be a promising strategy to describe grounding in a way that leaves open the question of whether or not the "*relata*" are possibly identical. Issues of identity should be settled in the light of a study of the *relata*, rather than by a conceptual decision. Thus, Jenkins seems to be right; her argument that paradigmatic cases of grounding relations can be given a reductive interpretation is persuasive. In the light of a comparison between the respective job-descriptions for grounding and reduction, this seems to be quite natural.

So, *prima facie*, a number of, but not all cases that seem to be adequately describable in terms of grounding are conceptually compatible with or even perfectly match a reductionist interpretation. If so, then grounding comes in two forms, in a reductionist and in a non-reductionist form. The (more general) grounding relation is then *underdetermined* with respect to the issue of identity.

Let me point to one alternative job description for grounding, a description which would clearly rule out a reductive interpretation. If the study of grounding is the study of *dependencies among different metaphysical layers of reality*, where talk of 'layers' does not allow for cross-layer identity, then reduction is not a case of grounding (and what merely appears to be *grounded* may turn out to *reduce* instead). Moreover, some theorists might want to argue that whenever we have diversity, directionality, and committal unity, we do not have identity; this is a substantive issue, and not a terminological one. I hope that the present book's first part has shed light on how we can conceive of a coherent explication of reduction that turns out to be compatible with diversity, directionality, and committal unity based on identity. On a liberal understanding, the grounding relation (or the way philosophers employ the term 'ground') is more encompassing: It reconciles diversity and directionality with metaphysically committal unity, either by identification *or* by dependence.[9]

The grounding debate is one of the last in which true armchair philosophy seems to flourish. Let us now turn to a final application of the explication of reduction, an application to interventionist models of causation and dependence. This debate comprises two aspects, at least. First, it aims at a counterfactual

[9]However, whereas there may be an interesting *a priori* route to metaphysical theories about the relation between a cheese and its wholes, particularized properties and their bearers, natural properties and moral or aesthetic properties, and sets and their elements, there is no such route to a theory about the relation between the mental and the physical; this seems to be a purely *empirical* matter.

definition of causation in terms of intervention. Secondly, it has been referred to in order to illuminate a pragmatic aspect of the scientific relevance of mechanistic explanations (Craver 2006). In the next and final section, I will briefly argue that reductive explanation does, once achieved, form a perfect epistemic basis for interventions.

9.4 Reduction and Intervention

Some interventions on a dependence-base result in a change in the dependent object. The following brief reflection on the notion of an intervention will enable us to adequately describe the connection between reduction and the pragmatically relevant aspect of interventions.

The literature on the pragmatics as well as the theoretical relevance of interventions in the philosophy of science focused on causal and mechanistic contexts, where it is taken for granted that the benefits of interventions stem from the (theoretically and pragmatically relevant) fact that when intervening on an object (or a variable in a representation of a causal system), we thereby change objects (or values of variables) that causally or mechanistically depend upon the object we intervened upon.[10]

In the debate on mechanisms, reference to interventions is frequent when it comes to distinguishing between mechanistic explanations and mere models of mechanisms that do not adequately capture the target entity's organization (*cf.* Craver 2006). Thus, it is introduced as an epistemological criterion to distinguish between mechanistic (or correct mechanistic) and non-mechanistic (or incorrect mechanistic) explanations. Moreover, it plays a role when it comes to judging issues regarding top-down "causation". The idea is that for there to be top-down "causes", we should find an intervention on the overall-mechanism that somehow alters its organization (cf. Craver 2007). These discussions do not directly bear upon reductive explanation, since they seem to presuppose that there are *two* objects involved in the intervention, one that is dependent and another one that forms the dependence base. However, necessarily, if x reduces to y, then an intervention on x is an intervention on y. This is trivially true, because (i) for x to reduce to y, $x = y$, and (ii) 'intervenes on' is extensional. Note that this changes once we take interventions to be interventions on variables in a representation of a system: The intervention on one variable results in a change in another variable does not ensure that this

[10]The vast literature on this topic mainly refers to Pearl (2000) and Woodward (2003). I assume that the reader is familiar with the basics. The task of finding definitions for causation will not be our concern here. Similarly, the role interventions may play in the discovery of reductions will not be addressed. If they do, the causal graphs that represent a system should be neutral with respect to which dependence relations they represent (see also footnote 11).

dependence mimics a dependence between two objects in the target system – the might just be *one* represented by *both*.

Thus, we have to look elsewhere for the connection between reduction and intervention. One interesting thought, pertinent, for example, in Woodward (2003) and Craver, is this: A *deeper* explanation enables us, at least in principle, to answer questions of the following form: 'What if things had been different?' This is supposed to capture the following connection: If we are familiar with the causal architecture, or the nature, or a constitutive property structure of an object, we can manipulate things *in an informed way*. This suggests a tight connection between reductive explanation and the pragmatic relevance of scientific interventions. In order to make the connection between interventions and reduction transparent, we need an idea of how interventions may become description relative. Once this idea is made precise, the connection between reduction and interventions can easily be explicated.

Interventions are description-dependent insofar as they are *intended* actions that are based on specific explicit representations of the object we intervene on. Necessarily, any intervention on water is an intervention on H_2O, and *vice versa*. However, an *intended* intervention can, intuitively, be construed as an intervention that is relative to a certain representation of the object intervened on. Let an *intended intervention* be an intervention on x that is planned with respect to a representation by a property structure that presents us with x. The hyper-intensionality is here inherited from the hyper-intensionality of intentionality. If I intend to intervene upon the temperature of a piece of ice, I do not thereby intend to intervene upon the kinetic energy and the lattice structures of the sum of corresponding H_2O molecules. Intended interventions are always description-, and, hence, property-structure-relative.[11]

Now, the relevance of reductive explanations for scientific interventions becomes transparent: Once we *learn* that an object x, when presented under a property structure *PS1*, reduces to x, when presented under a property structure *PS2*, we get conceptual access to a number of intended interventions with respect to *PS2*. Assume that we learn that, say, the occurrence of fever is a complex event, involving the release of PGE2 which, acting on the hypothalamus, causes the body temperature to raise. Acquiring this knowledge, and acquiring the knowledge that fever occurs in virtue of the occurrence of this mechanism, we are in a position

[11]Note that interventionist models of causation talk about representation-dependence as well. However, it is noteworthy that here, representation dependence concerns the different causes that are presented in a causal graph, rather than the different ways one and the same cause may be presented in a causal graph. A possible problem for interventionist definitions of causation is that they cannot properly distinguish between causation, on the one hand, and other dependence relations on the other. Unless we stipulate that in a causal graph, any variable has to represent another entity, and that synchronic dependence relations are excluded by fiat, they seem to be highly problematic (similar worries have been risen in the discussion of how interventionism relates to the causal exclusion argument and, hence, to epiphenomenalism; *cf.* Baumgartner (2010) for a critical examination of this discussion).

to plan intended interventions on the mechanism. Note that this does not come for free with the truth of an identity statement: Assume that fever = the medical sign which is most often mistaken for a disease (in the actual world). Knowledge of this identity statement does not enable us to intervene on fever. Why is this so? One may suggest that this is so *because* being a fever does not *depend* on being the medical sign that is most often mistaken for a disease (in the actual world). Reduction is a form of dependence. Knowledge of reduction statements puts one in the position to plan intended interventions one might not have access to otherwise. Interestingly, such knowledge also enables us to get access to intended interventions with respect to the constitutive property structure, which may have an effect on the properties under which the same entity is presented at the reduced level. Assume that the folk-concept of fever corresponds to a property structure that represents fever in terms of its symptoms. Once we come to know what fever is on a physiological level – in terms of a physiological property-structure –, we learn how to intervene upon it by designing effective drugs. These drugs enable us to intervene upon the symptoms by intentionally intervening upon the physiology. Thus, reductive explanation connects to intentional intervention as follows: If a subject *s* knows that *x*, when presented under *PS1*, reduces to *x*, when presented under *PS2*, then *s* is (at least under favorable circumstances)[12] in a position to intentionally intervene upon *x* with respect to *PS1* by intentionally intervening on *x* with respect to *PS2*.

Since at the reducing level, property structures give access to *constitutive* property structures, knowledge of a truth about a reduction relation increases, at least in principle, the set of possible intended interventions on that object. Thus, reductive explanations play an important part in the pragmatically relevant outcome of scientific enterprises. Even a partially correct reductive explanation would suffice: As long as part of the low-level representation – say, fever under some of its physiological properties – is correct, an intervention along the lines just described may succeed. Correspondingly, a partial reduction (i.e. where a whole is not fully re-described in terms of its constituents) will be sufficient. This is not sufficient for full-blown reduction. But it may be sufficient for full-blown intended interventions.

9.5 Conclusion

In this chapter, it was shown that the concept explicated in the first part of this book can be fruitfully applied to topics outside the reduction debate, and the argument proposed in Chap. 8, that the account proposed here is more fundamental than its

[12]That is: The intervention must in principle be possible and the subject needs the relevant means and skills. Knowledge itself does not guarantee the possibility of an intentional intervention. However, it is required for this kind of intervention. Moreover, one should, maybe, not stress the notion of knowledge to much in this context. Even true believe might be sufficient.

holistic rivals, has been continued. The answer to the guiding question of this second part of the book has thus been completed:

Q2:	How can this explication be further motivated?
...	
Th. 7:	It is as committal as and more fundamental than rival explications.
Th. 8:	It sheds light on closely related issues, such as reduction and unification, pragmatic benefits of reduction, and notions of scientific levels.

This chapter continued the debate on the relation between holistic approaches to reduction (Chap. 8), with an eye on the relation between reduction and forms of unification, and it applied the explication proposed here to issues such as reduction and grounding, reduction and intervention, reduction and physicalism, and reduction and the notion of scientific levels. One notion of a scientific level is tied to modes of presentations, in a way that can, again, be illuminated referring to property structures. Unification is to be expected in reductions: We subsume apparently different phenomena under one coherent conceptual scheme, achieve ontological unification and epistemic as well as explanatory unification. Reduction is a cognate, or a version of grounding, and knowledge of true reduction statements has an impact on the set of intended interventions we have access to.

Bibliography

Baumgartner, Michael. 2010. Interventionism and epiphenomenalism. *Canadian Journal of Philosophy* 40: 359–384.

Beckermann, Ansgar. 2009. What is property physicalism? In *The Oxford handbook of philosophy of mind*, ed. Sven Walter, Brian P. McLaughlin, and Ansgar Beckermann, 152–172. Oxford: Oxford University Press.

Braddon-Mitchell, David, and Frank Jackson. 1996. *Philosophy of mind and cognition*. Oxford: Blackwell Publishing.

Cartwright, Nancy. 1983. *How the laws of physics lie*. Oxford: Oxford University Press.

Cartwright, Nancy. 1989. *Nature's capacities and their measurement*. Oxford: Oxford University Press.

Causey, Robert. 1972. Attribute identities in microreductions. *Journal of Philosophy* 64: 407–422.

Causey, Robert. 1977. *Unity of science*. Dordrecht: Reidel.

Chalmers, David. 1996. *The conscious mind*. Oxford: Oxford University Press.

Correia, Fabrice. 2005. *Existential dependence and cognate notions*. München: Philosophia.

Correia, Fabrice. 2008. Ontological dependence. *Philosophy Compass* 3(5): 1013–1032.

Craver, Carl. 2006. When mechanistic models explain. *Synthese* 153: 355–376.

Craver, Carl. 2007. *Explaining the brain: Mechanisms and the mosaic unity of neuroscience*. Oxford: Oxford University Press/Clarendon Press.

Cussins, Adrian. 1992. The limitations of pluralism. In *Reduction, explanation, and realism*, ed. Kathleen Lennon and David Charles, 179–224. Oxford: Oxford University Press.

Darden, Lindley, and Nancy Maull. 1977. Interfield theories. *Philosophy of Science* 44: 43–64.

Dupré, John. 1993. *The disorder of things. Metaphysical foundations of the disunity of science*. Cambridge, MA: Harvard University Press.

Fine, Kit. 2001. The question of realism. *Philosophers' Imprint* 1: 1–30.

Fine, Kit. 2005. *Modality and tense. Philosophical papers*. Oxford: Oxford University Press.

Fodor, Jerry. 1974. Special sciences: Or the disunity of science as a working hypothesis. *Synthese* 28: 97–115.
Friedman, Michael. 1974. Explanation and scientific understanding. *Journal of Philosophy* 71: 5–19.
Friedman, Kenneth. 1982. Is intertheoretic reduction feasible? *The British Journal for the Philosophy of Science* 33: 17–40.
Hempel, Carl G. 1969. Reduction: Ontological and linguistic facets. In *Philosophy, science, and method: Essays in honor of Ernest Nagel*, ed. Morton White, Sidney Morgenbesser, and Patrick Suppes, 179–199. New York: St. Martin's Press.
Jenkins, Carrie S. 2011. Is metaphysical dependence irreflexive? *The Monist* 94: 267–276.
Kemeny, John G., and Paul Oppenheim. 1956. On reduction. *Philosophical Studies* 7: 6–19.
Kitcher, Philip. 1981. Explanatory unification. *Philosophy of Science* 48: 507–531.
Kitcher, Philip. 1989. Explanatory unification and the causal structure of the world. In *Scientific explanation*, ed. Wesley Salmon and Philip Kitcher, 410–505. Minneapolis: University of Minnesota Press.
Lennon, Kathleen. 1992. Reduction, causality, and normativity. In *Reduction, explanation, and realism*, ed. Kathleen Lennon and David Charles, 225–238. Oxford: Clarendon.
Lewis, David. 1970. Nominalistic set theory. *Noûs* 4: 225–240.
Meehl, Paul E., and Wilfrid Sellars. 1956. The concept of emergence. In *The foundations of science and the concepts of psychology and psychoanalysis*, ed. Michael Scriven and Herbert Feigl, 239–252. Minneapolis: University of Minnesota Press.
Nagel, Ernest. 1961. *The structure of science. Problems in the logic of explanation*. New York: Harcourt, Brace & World, Inc.
Papineau, David. 2007. Phenomenal and perceptual concepts. In *Phenomenal concepts and phenomenal knowledge*, ed. Torin Alter and Sven Walter, 111–144. Oxford: Oxford University Press.
Pearl, Judea. 2000. *Causality: Models, reasoning, and inference*. Cambridge: Cambridge University Press.
Place, Ullin. 1956. Is consciousness a brain process. *British Journal of Psychology* 47: 44–50.
Place, Ullin. 1960. Materialism as a scientific hypothesis. *Philosophical Review* 69: 101–104.
Schaffer, Jonathan. 2009. On what grounds what. In *Metametaphysics: New essays on the foundations of ontology*, ed. David Chalmers, David Manley, and Ryan Wasserman, 347–383. Oxford: Oxford University Press.
Simons, Peter. 1987. *Parts*. Oxford: Clarendon.
Smart, John J.C. 1959. Sensations and brain processes. *Philosophical Review* 68: 141–156.
Smart, John J.C. 1978. The content of physicalism. *Philosophical Quarterly* 28: 339–341.
Sober, Elliot. 1999. The multiple realizability argument against reductionism. *Philosophy of Science* 66: 542–564.
Trogdon, Kelly. 2013. An introduction to grounding. In *Varieties of dependence*, ed. Miguel Hoeltje, Benjamin Schnieder, and Alexander Steinberg, 97–122. München: Philosophia Verlag.
Van Gulick, Robert N. 1992. Nonreductive materialism and the nature of intertheoretical constraint. In *Emergence or reduction? essays on the prospects of non-reductive physicalism*, ed. Ansgar Beckermann, Hans Flohr, and Jaegwon Kim, 157–179. Berlin: de Gruyter.
Wimsatt, William C. 1976. Reductive explanation: A functional account. In *PSA: Proceedings of the Biennial meeting of the Philosophy of Science Association, 1974*, ed. G. Pearce, A.C. Michalos, C.A. Hooker, and R.S. Cohen, 671–710. Dordrecht: Reidel.
Woodward, James. 2003. *Making things happen. A theory of causal explanation*. Oxford: Oxford University Press.

Chapter 10
The Reductionist's Commitment

Reductionism is the doctrine that we live in a metaphysically unified world, the constituents of which can be presented by various sorts of modes of presentation. Chapter 3 opened the discussion, offering a solution to the puzzle that arises form the slogan that reduction reconciles diversity and directionality with strong unity. This solution suggests that reduction is sensitive to the conceptual contents under which the reducing or reduced object is presented in a true reduction statement. This suggests that diversity is conceptual in nature. Chapter 4 argued that directionality is to be accounted for in terms of reductive explanation, a cognate of mechanistic explanation. Building on the results of Chaps. 3 and 4, Chap. 5 offered an explication of a core notion of reduction in the sense of 'explication' described in Chap. 2. The explication is motivated as follows: It captures the *paradigmatic cases* as well as the *slogan* that reduction reconciles diversity and directionality with strong unity, without relying on elimination. Moreover, to the extent that the explication reflects the results of the discussions of Chaps. 3 and 4, it seems justified. It offers a way to solve the puzzle and to make sense of reductive explanation as a cognate of mechanistic explanation.

The second part of the book offered further motivation for endorsing the explication. Chapter 6 argued that the explication captures and illuminates the use of 'reduction' and its cognates in large parts of the philosophy of mind. However, one may doubt that it does justice to the use of 'reduction' in the philosophy of science. In particular, one may worry that (i) at best, a notion of identity-based reduction, even when construed as an inter-theory relation, plays a minor role in the philosophy of science, and (ii) holistic notions of theory reduction provide the means to deal with the alleged problems all by themselves. Moreover, it has been assumed that notions of theory reduction are to be preferred because they are (iii) less committal, and (iv) more fundamental than the proposal offered here. Chapter 7 argues that (i) is mistaken, thereby paving the way for a discussion of the criticisms expressed by (ii)–(iv) in Chap. 8 and, partly, in Chap. 9. It was argued that identity-based reduction is crucial for an appropriate understanding of models of reduction and replacement in the philosophy of science (against (i)), that there are serious doubts that models of theory reduction provide the means to deal with the problems

R. van Riel, *The Concept of Reduction*, Philosophical Studies Series 121,
DOI 10.1007/978-3-319-04162-9_10,

discussed here (against (ii)), that it is not the case that these characterizations are less committal – rather, they are equally expensive (against (iii)) – and that the notion explicated here is more fundamental than holistic notions of reduction (against (iv)). In addition, the final Chap. 9 discussed the relevance of the so explicated notion for topics other than those that are directly concerned with reduction, thereby illustrating the fruitfulness of the explication in other contexts.

Thus, the explication captures an important use of the term, it meets the intuitive description and paradigmatic cases, it can deal with alleged problems, it turns out to be more fundamental than its most prominent rival, and it is fruitful not only in the context of reduction debates.

There is a promising way to reconcile descriptive or conceptual diversity with explanatory directionality and unity that is based on identification. On the interpretation of the concept of reduction proposed here, reductionism is committed to the idea that what appear to be different layers of reality are in fact different layers of modes of presentation of reality. Explanatory dependencies among these different levels organize the various reductive hierarchies. The sensitivity to modes of presentation is reflected by conceptions of scientific levels, which inspired a characterization of two notions of physicalism. Reconciling diversity and directionality with strong unity in this way enables us to account for epistemic features of alleged cases of reductions, and ontological, epistemic and explanatory unification are to be expected in reductions. Similarly, once a reduction of one theory to another is achieved, we will come to see that the reducing theory explains the phenomena of the reduced theory, and we gain the resources to explain why the reduced theory worked as well as it did. We can, in the light of the so explicated notion, explain how reduction relates to the pragmatic value of expanding the range of possible intended interventions. The notion of "ontological" reduction proposed here is, in this sense, prior to notions of theory reduction, which tie reductions to holistic features of theories.

Reductionists are committed to the idea that reduction-relations structure the actual world. The explication proposed here gives a coherent picture of what these relations consist in, and it performs better than its rivals. The reductionist's commitment has thus been made explicit.

MIX
Papier aus verantwortungsvollen Quellen
Paper from responsible sources
FSC® C105338

If you have any concerns about our products,
you can contact us on
ProductSafety@springernature.com

In case Publisher is established outside the EU,
the EU authorized representative is:
Springer Nature Customer Service Center GmbH
Europaplatz 3, 69115 Heidelberg, Germany

Printed by Libri Plureos GmbH
in Hamburg, Germany